Holt Mathematics

Chapter 4 Resource Book

HOLT, RINEHART AND WINSTON

A Harcourt Education Company

Orlando • **Austin** • New York • San Diego • London

ISBN 0-03-078299-6

6 7   170   09 08

CONTENTS

Blackline Masters

Holt Mathematics

Holt Mathematics

Date ______________

Dear Family,

In this chapter, your child will learn about patterns, functions, and graphs. Your child will learn that the same set of data can be represented in different ways, including tables, equations, and graphs. Your child will also identify patterns in data, interpret sets of data, and use equations to represent relationships among sets of data.

Your child will begin by graphing **ordered pairs** on a **coordinate plane**.

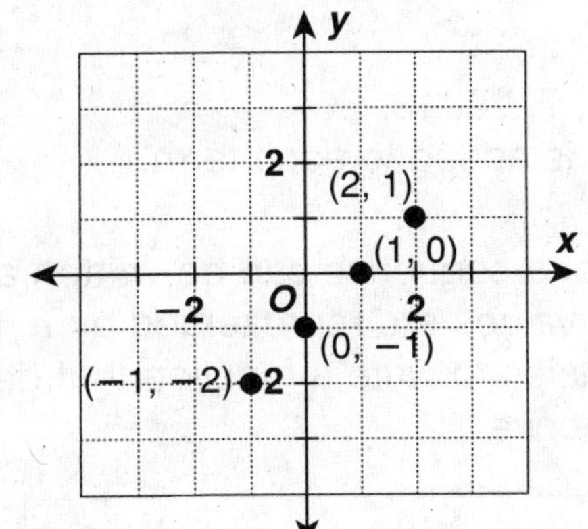

Ordered Pairs

x	y
−1	−2
0	−1
1	0
2	1

(x, y)
(−1, −2)
(0, −1)
(1, 0)
(2, 1)

Next, your child will relate graphs to situations. He or she will read about situations and will identify or draw graphs that match these situations. For example, this graph shows a situation in which the ink in a computer cartridge is completely used up. Then the cartridge is refilled, and the ink is used up again.

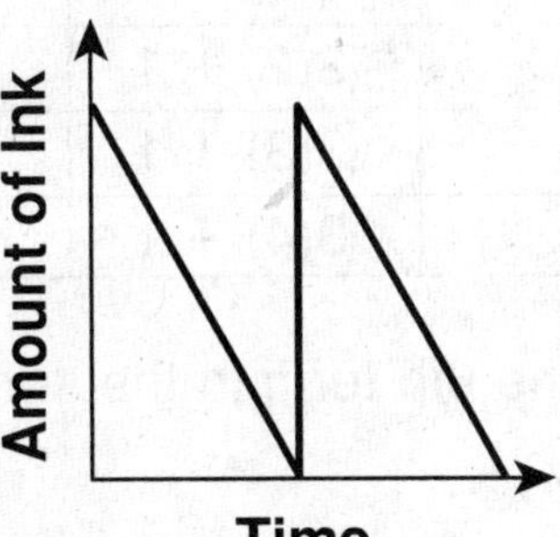

Your child will then learn about **functions** and will graph them. A mathematical function is a rule that pairs one output value to each input value. For example, the rule might be "multiply a number by 2 and then subtract 1 from the result." The equation that represents this function is $y = 2x - 1$, where x is the input and y is the output. You can use a table to organize input and output values and write ordered pairs which can be graphed.

Input Pair	Rule	Output	Ordered
x	2x − 1	y	(x, y)
−1	2(−1) − 1	−3	(−1, −3)
0	2(0) − 1	−1	(0, −1)

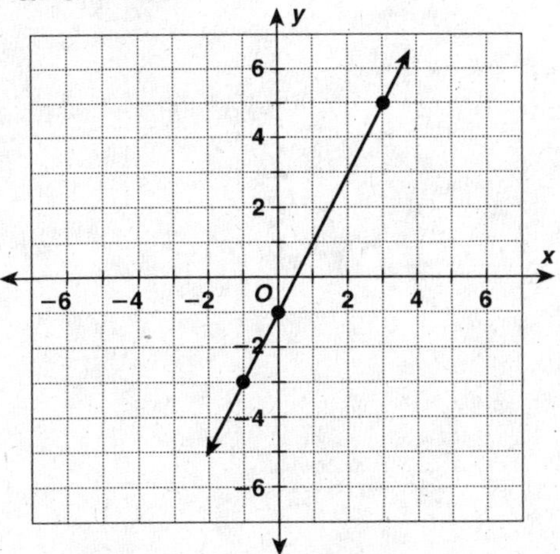

1

Holt Mathematics

The function $y = 4x - 2$ is a **linear function** because its graph is a nonvertical line. A linear function describes an output that is changing at a constant rate. Your child will learn to use functions to solve problems involving constant rates of change, such as rising temperatures.

Your child will also look for patterns in **sequences**.

In an **arithmetic sequence**, the same number is added to each term to get the next term.

Example: 4, 7, 10, 13, …
          3 is added to each term.

In a **geometric sequence**, each term is multiplied by the same number to get the next term.

Example: 2, 8, 32, 128, …
          Each term is multiplied by 4.

The terms of a sequence can be written as outputs (y) in a function table. Input values are represented by n, the number of each term. This table helps to write a function that describes the arithmetic sequence above.

n	Rule	y
1	$3(1) + 1$	4
2	$3(2) + 1$	7
3	$3(3) + 1$	10
4	$3(4) + 1$	13

The rule is to multiply n by 3 and add 1. The function is $y = 3n + 1$.

To find the 6th term in the sequence, substitute 6 for n in the function.

$y = 3n + 1$

$y = 3(6) + 1$

$y = 19$

The 6th term is 19.

For additional resources, visit go.hrw.com and enter the keyword MS7 Parent.

Holt Mathematics

Practice A

The Coordinate Plane

Identify the quadrant of each point.

1. *A* _______________

2. *B* _______________

3. *C* _______________

4. *D* _______________

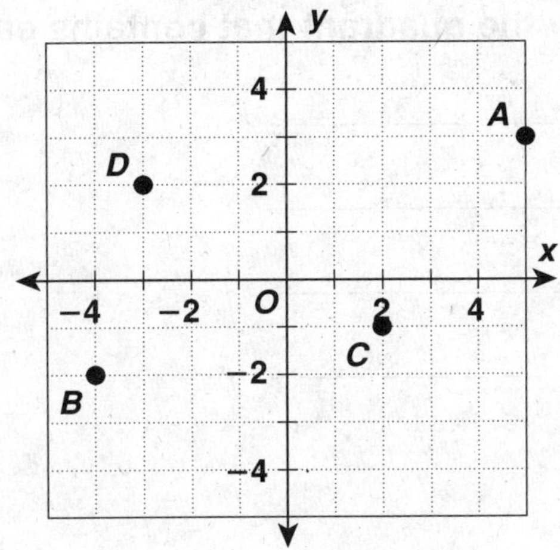

Plot each point.

5. (0, 3)

6. (−2, 4)

7. (5, 1)

8. (−4, −3)

9. (3, −2)

10. (−3, −1)

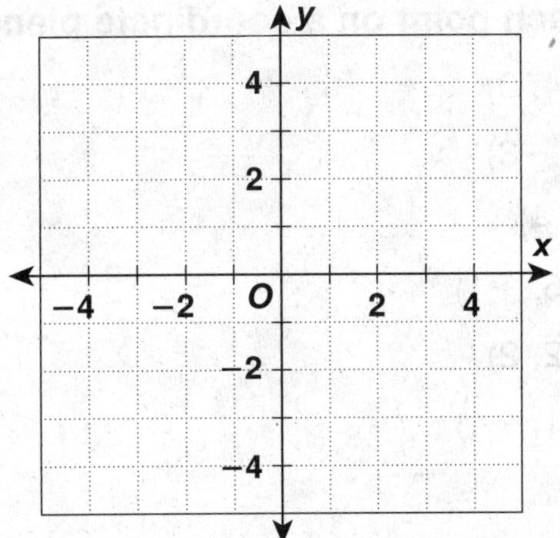

Give the coordinates of each point.

11. *J* _______________

12. *K* _______________

13. *L* _______________

14. *M* _______________

15. *N* _______________

16. *P* _______________

17. *R* _______________

18. *S* _______________

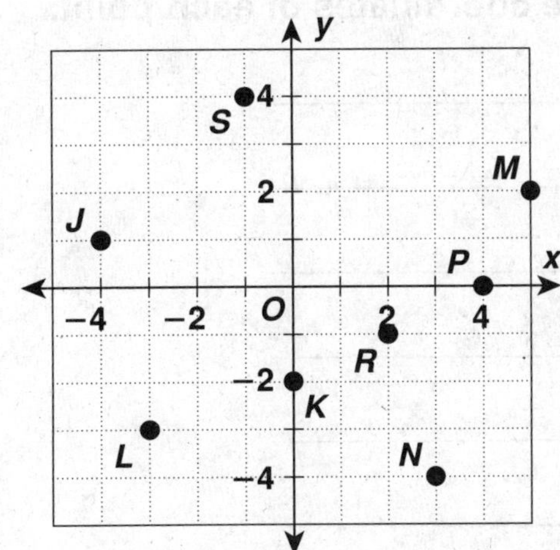

Holt Mathematics

Practice B
The Coordinate Plane

Identify the quadrant that contains each point.

1. A ____________________

2. B ____________________

3. C ____________________

4. D ____________________

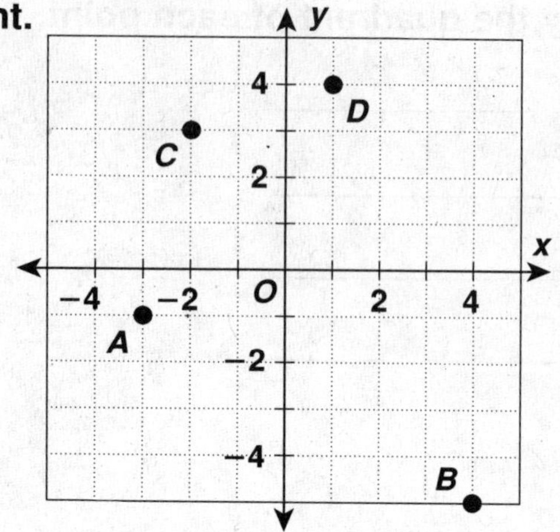

Plot each point on a coordinate plane.

5. $(-4, 0)$

6. $(3, -3)$

7. $(1, 4)$

8. $(-5, -1)$

9. $(-2, 2)$

10. $(-1, -4)$

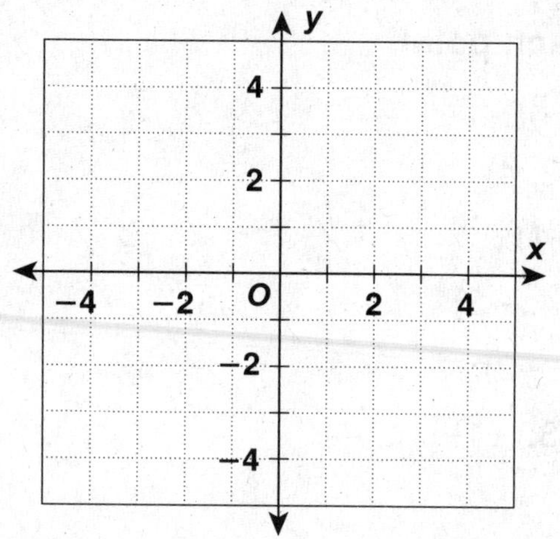

Give the coordinates of each point.

11. P ____________________

12. Q ____________________

13. R ____________________

14. S ____________________

15. T ____________________

16. U ____________________

17. W ____________________

18. X ____________________

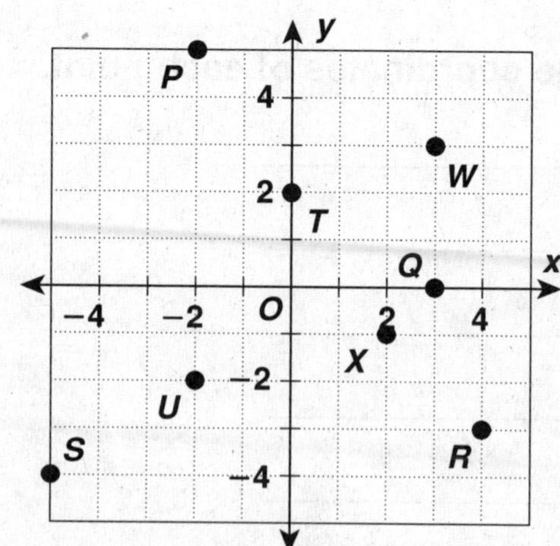

Holt Mathematics

Practice C

The Coordinate Plane

Identify the quadrant that contains each point.
Give the coordinates of each point.

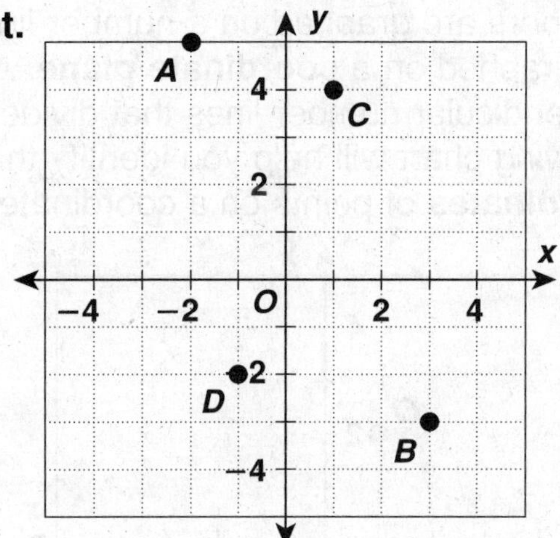

1. *A* _______________

2. *B* _______________

3. *C* _______________

4. *D* _______________

Plot each point on a coordinate plane.

5. $(-2, -2)$

6. $(4, -5)$

7. $(0, -2)$

8. $(-4, 4)$

9. $(3, 0)$

10. $(-3, 1)$

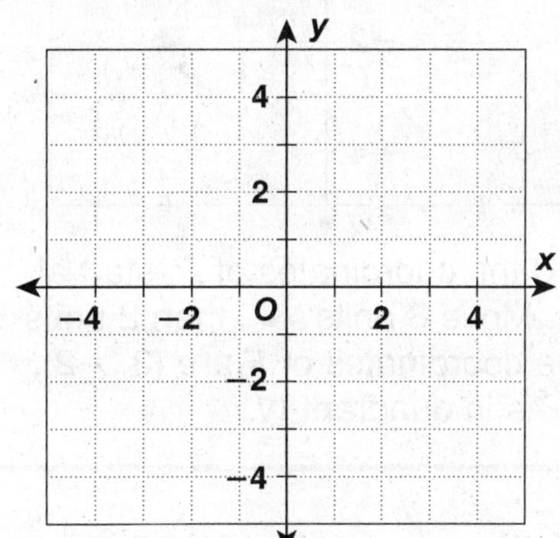

11. Graph the points $(3, -2)$, $(-3, -2)$ $(-4, 3)$ and $(2, 3)$. Connect each point in the order listed. Name the figure and the quadrants in which it is located.

____

____

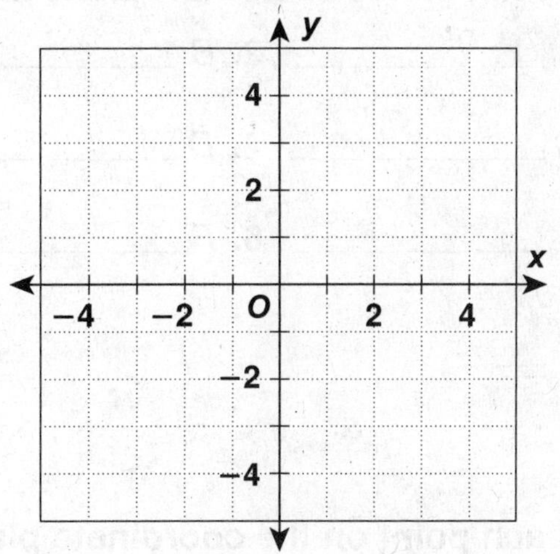

Holt Mathematics

LESSON 4-1 Reteach
The Coordinate Plane

Numbers are graphed on a number line. **Ordered pairs** of numbers are graphed on a **coordinate plane.** A coordinate plane has two perpendicular number lines that divide it into **4 quadrants.** The following chart will help you identify the quadrants and the **coordinates** of points on a coordinate plane.

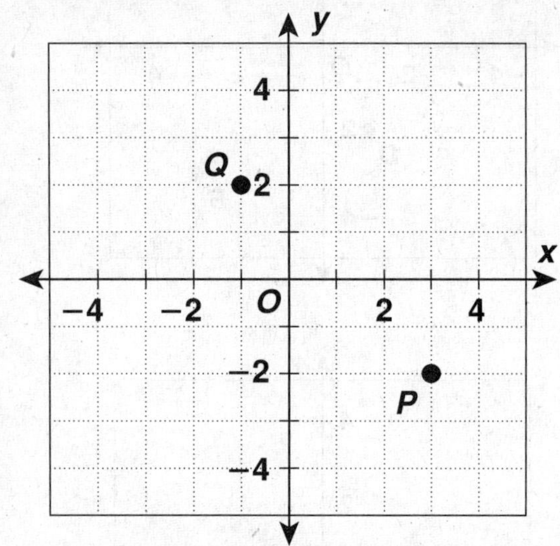

Quadrant II	Quadrant I
$(-,+)$	$(+,+)$
$(\leftarrow, \uparrow)$	$(\rightarrow, \uparrow)$
Quadrant III	**Quadrant IV**
$(-,-)$	$(+,-)$
$(\leftarrow, \downarrow)$	$(\rightarrow, \downarrow)$

To find the coordinates of P, start at $(0, 0)$. Move 3 units $\rightarrow$, then 2 units $\downarrow$. So the coordinates of P are $(3, -2)$, and P is in quadrant IV.

To plot point Q with coordinates $(-1, 2)$, start at $(0, 0)$. Move 1 unit $\leftarrow$, then 2 units $\uparrow$. Q is in quadrant II.

Identify the quadrant and the coordinates of each point on the coordinate plane at the right.

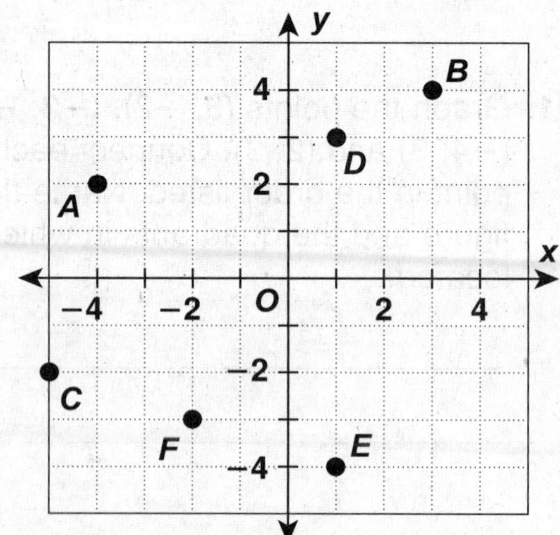

1. A _____________   2. B _____________

3. C _____________   4. D _____________

5. E _____________   6. F _____________

Plot each point on the coordinate plane above.

7. $G\,(-4, -3)$ 8. $H\,(0, -2)$ 9. $J\,(3, -5)$ 10. $K\,(-3, 1)$

11. $L\,(4, -1)$ 12. $M\,(-3, 4)$ 13. $N\,(-1, 3)$ 14. $Z\,(3, 0)$

6

Holt Mathematics

LESSON 4-1 Challenge

Where in the World?

Graph each point on the grid below. Connect each point to the previous one as you graph it. Then connect the last point to the first point.

1. $(0, -10)$ **2.** $(-1, -9)$ **3.** $(-2.5, -7)$ **4.** $(-5, -7)$

5. $(-6, -5)$ **6.** $(-10, -5)$ **7.** $(-13, -3)$ **8.** $(-15, -1)$

9. $(-16, 2)$ **10.** $(-15, 8)$ **11.** $(-15, 10)$ **12.** $(-3, 9)$

13. $(4, 8)$ **14.** $(4, 7)$ **15.** $(6, 8)$ **16.** $(6, 4)$

17. $(8, 6)$ **18.** $(9, 6)$ **19.** $(9, 3)$ **20.** $(11, 5)$

21. $(16, 10)$ **22.** $(18, 8)$ **23.** $(16, 6)$ **24.** $(18, 4)$

25. $(14, 1)$ **26.** $(14, -1)$ **27.** $(11, -5)$ **28.** $(12.5, -8)$

29. $(13, -10)$ **30.** $(11, -9)$ **31.** $(9, -6)$ **32.** $(2.5, -7)$

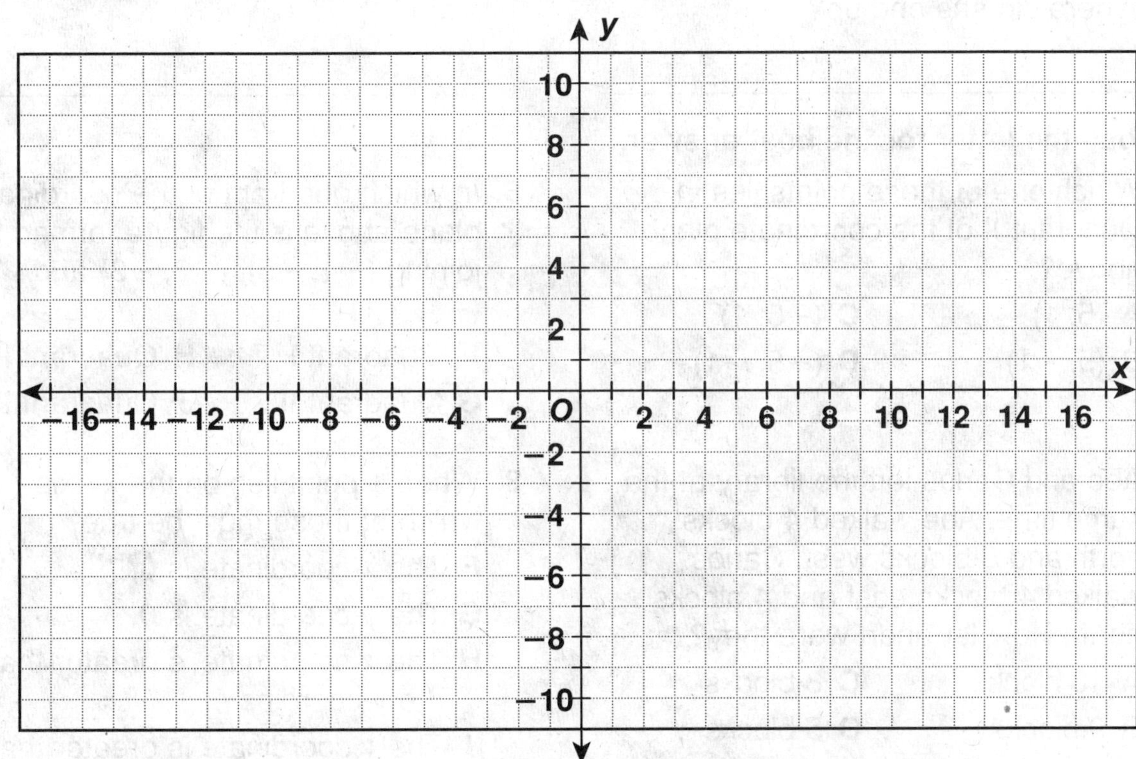

33. In which state is the point $(0, -8)$? ________________________

34. Name a point in the state of Florida. ________________________

Holt Mathematics

<table><tr><td>**LESSON**
4-1</td><td></td></tr></table>

Problem Solving
The Coordinate Plane

Write the correct answer.

1. Use the coordinate plane at right. In which quadrant(s) would the figure drawn by connecting points *J, K,* and *N* be?

2. Use the coordinate plane at right. In which quadrant(s) would the figure drawn by connecting points *C, F,* and *M* be?

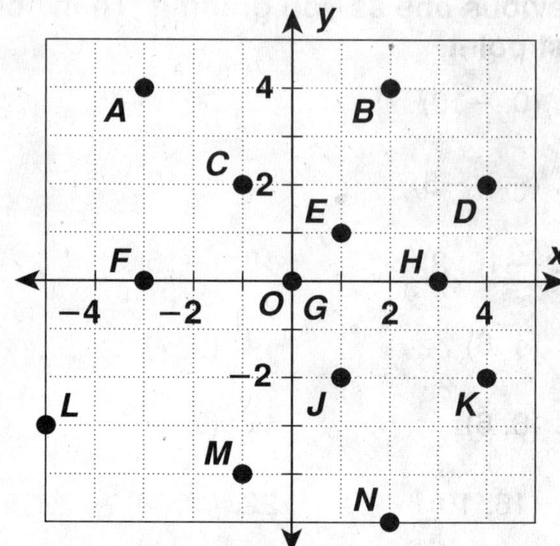

3. Maxine left home and walked 5 blocks north, 5 blocks west, 5 blocks south, and 5 blocks east. Where did she end up?

4. Mr. Chin drove 2 miles north, then 3 miles east, then 2 miles south. How far is Mr. Chin from where he started?

Choose the letter for the best answer.

5. Which one of these points lies in Quadrant II of the coordinate plane above?

   A (5, 1)   **C** (−5, 1)

   B (5, −1)   **D** (−5, −1)

6. In which quadrant of the coordinate plane above is the figure formed by joining (−4, −5), (−2, −3) and (−1, −1)?

   F Quadrant I   **H** Quadrant III

   G Quadrant II   **J** Quadrant IV

7. Abe and Carlos left the library at the same time. Abe walked 4 blocks north and 5 blocks west. Carlos walked 4 blocks east and 4 blocks north. How far apart were they?

   A 10 blocks   **C** 8 blocks

   B 9 blocks   **D** 5 blocks

8. When a point lies on the *x*-axis, which of these must be true?

   F The *x*-coordinate is 0.

   G The *y*-coordinate is 0.

   H The *x*-coordinate is greater than the *y*-coordinate.

   J The *y*-coordinate is greater than the *x*-coordinate.

Holt Mathematics

Name _________________________________ Date __________ Class __________

LESSON 4-1 **Reading Strategies**
Using Relevant Information

Horizontal number lines and vertical number lines form a grid called
the **coordinate plane.**

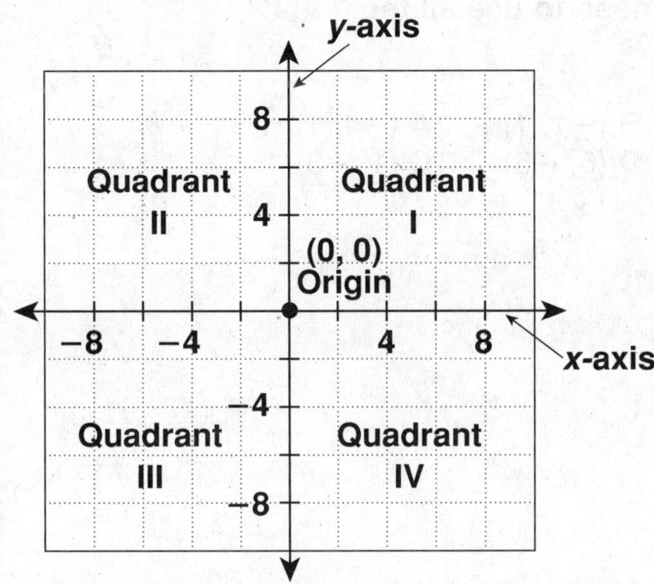

1. What is the name of the horizontal number line? _____________________

2. What is the name of the vertical number line? _____________________

3. The number lines meet at point (0, 0). What is that point called? _____________

4. What are the four parts that divide the coordinate plane called? _______________

To locate a point on the coordinate plane, you always start at the
origin. You first move either to the right or left along the **x-axis.**

**Write "positive" or "negative" to show which direction you are
moving from zero.**

5. If you move to the right, you are moving in a _________________ direction.

6. If you move to the left, you are moving in a _________________ direction.

From your position on the x-axis, you move up or down along the **y-axis.**

**Write "positive" or "negative" to show which direction you are
moving from zero.**

7. If you move up, you are moving in a _________________ direction.

8. If you move down, you are moving in a _________________ direction.

Holt Mathematics

LESSON 4-1 Puzzles, Twisters & Teasers
Plane Thinking!

Plot the points in the coordinate plane below. Fill in the blanks to form a sentence, then connect the points to make a picture that matches your sentence. You won't need to use all the letters in your sentence.

S (1, 5)   **N** (3, 1)   **T** (6, 1)   **R** (−1, 1)   **A** (−4, 1)
O (−1, −2)   **I** (3, −2)   **B** (−4, −5)   **D** (6, −5)   **C** (1, −3)

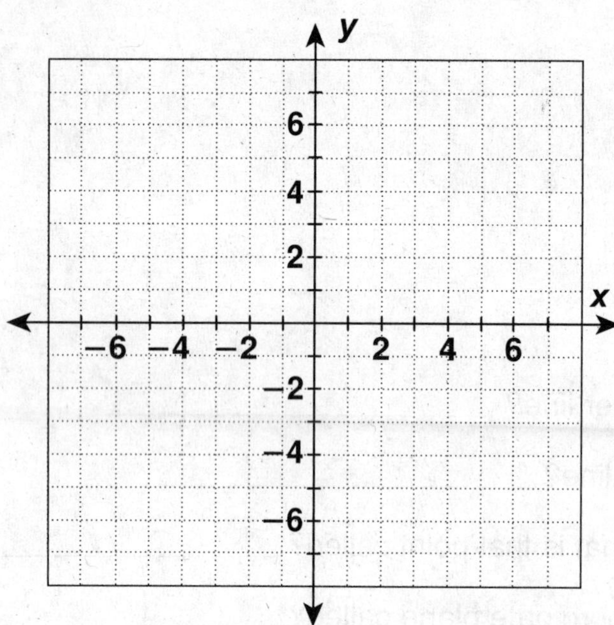

____ ____ ____ ____ ____   ____ ____
(−4, 1)   (1, 5) (6, 1) (−4, 1) (−1, 1)   (3, −2) (1, 5)

____ ____ ____ ____ !
(−4, −5) (−1, −2) (−1, 1)   (3, 1)

10   **Holt Mathematics**

LESSON 4-2	**Practice A**

Practice A
Tables and Graphs

Write each ordered pair from the table.

1.
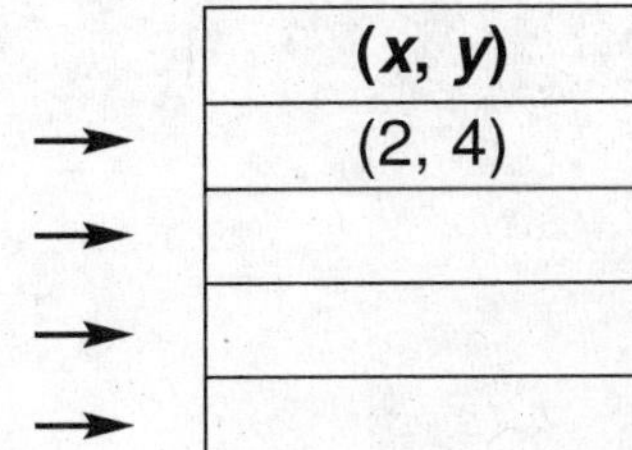

x	y
2	4
3	5
4	6
5	7

(x, y)
(2, 4)

Graph the ordered pairs from the table.

2.

x	y
−1	3
0	1
1	−1
2	−3

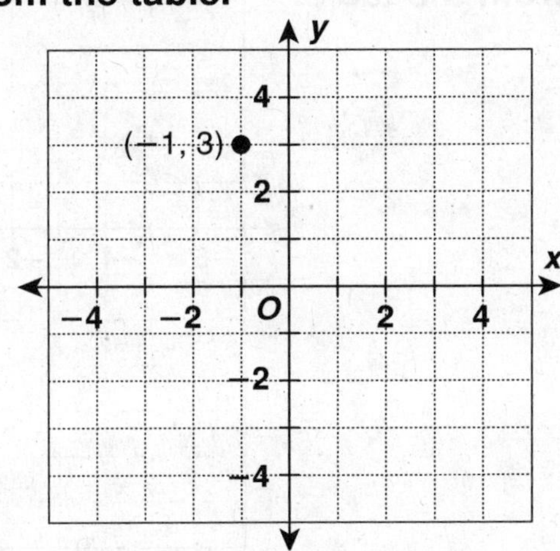

3. The table shows the cost for different weights of bananas.
   Graph the data.

Pounds of Bananas	Total Cost
2	$1
4	$2
6	$3
8	$4

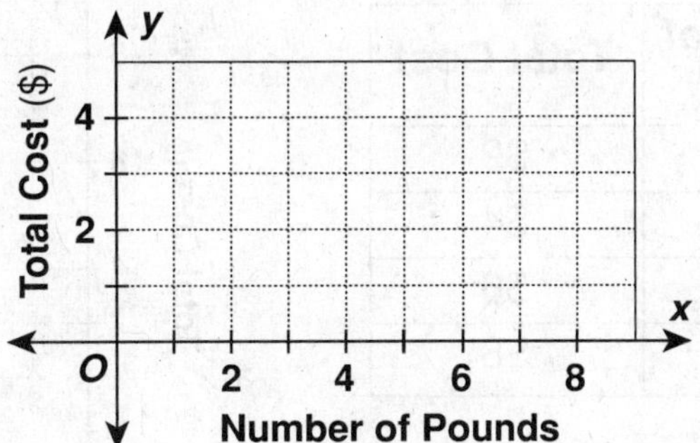

What appears to be the relationship between the number of
pounds and total cost?

____

11

Holt Mathematics

Practice B
Tables and Graphs

Write each ordered pair from the table.

1. 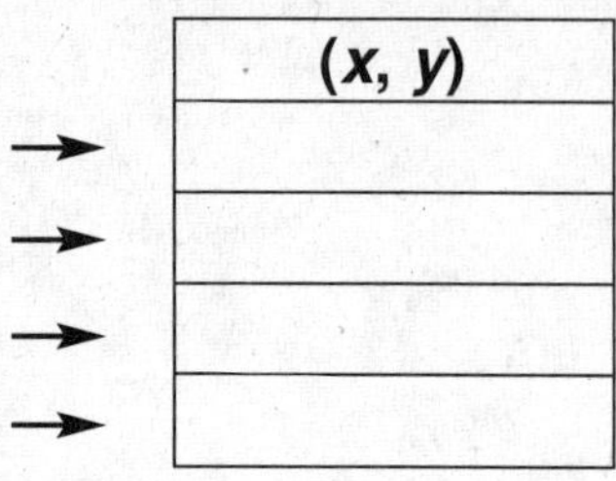

x	y
−12	−8
−9	−4
−6	0
−3	4

(x, y)

Graph the ordered pairs from the table.

2.

x	y
−5	4
−3	3
−1	2
1	1

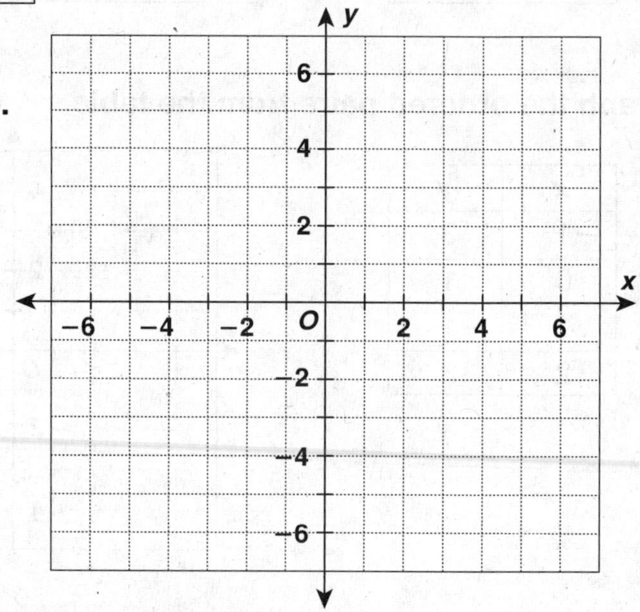

3. The table shows the cost of buying different numbers of souvenir pens. Graph the data.

Number of Pens	Total Cost
1	$2
2	$4
3	$6
4	$8

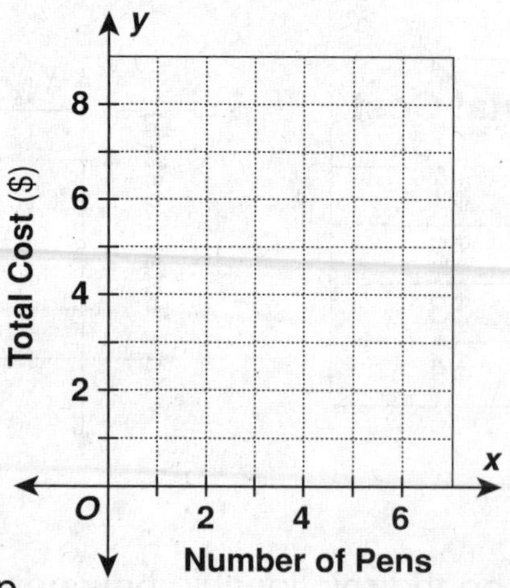

What appears to be the relationship between the number of pens and total cost?

____

Holt Mathematics

Practice C
Tables and Graphs

Write each ordered pair from the table.

1.

x	y
−7	12
−4	10
−1	8
2	6

(x, y)

Graph the ordered pairs from the table.

2.

x	y
−10	0
−5	−5
0	−10
5	−15

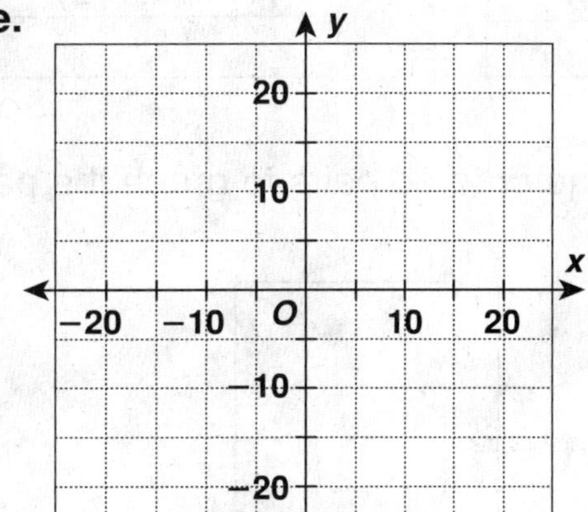

3. A 40-gallon water tank that was full developed a leak.

Elapsed Time (hr)	Water Remaining (gal)
0	40
1	35
2	30
3	25

a. Make a graph of the data.

b. Explain how you can use the graph to find the amount of water remaining after 6 hours.

____

Holt Mathematics

LESSON 4-2 Reteach
Tables and Graphs

You can write an ordered pair for each pair of values in a table of values.
Write the x value as the first number in the ordered pair.
Write the y value as the second number in the ordered pair.

Complete the missing values in the ordered pairs.

x	y		Ordered Pairs (x, y)
−2	−4	→	(−2, −4)
1. 0	−3	→	(0,)
2. 2	−2	→	(, −2)
3. 4	−1	→	(,)

You can use the ordered pairs to graph the pairs of values in the table.

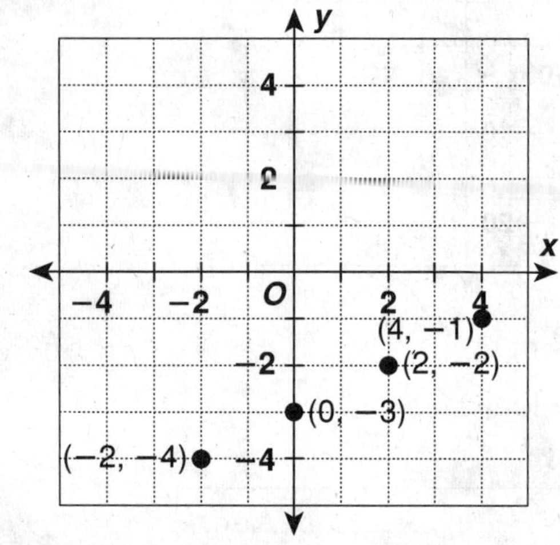

Write the ordered pairs from the table. Then graph the ordered pairs.

4.

x	y		(x, y)
−2	−1	→	
−1	0	→	
0	1	→	
1	2	→	

14

Holt Mathematics

Challenge
LESSON 4-2
Patterns on a Coordinate Plane

You can graph points and shapes on a coordinate plane. If you multiply the coordinates of these points by negative numbers, you will find patterns in the results.

1. Use the values from Table A to complete Table B.

2. Graph the ordered pairs from Table A. Connect each point to the previous point as you graph it. Then connect the last point to the first. Do the same for Table B.

3. How does multiplying the x-coordinates of a figure by −1 change the location of a figure?

____

____

Table A

x	y
1	3
2	1
3	2
4	4

Table B

−x	y
−1	3

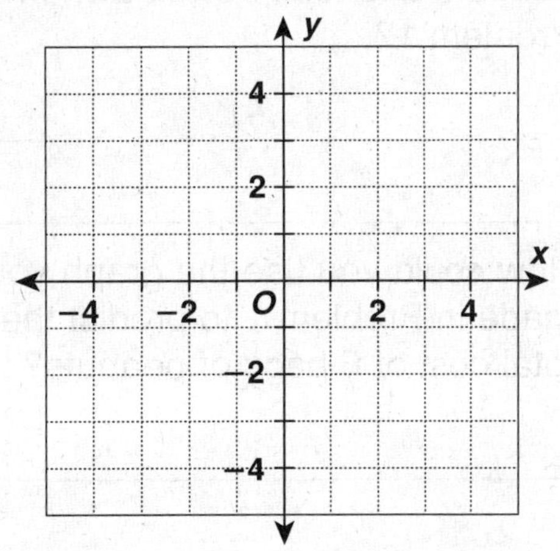

4. Use the values from Table C to complete Table D.

5. Graph the ordered pairs from Table C. Connect each point to the previous point as you graph it. Then connect the last point to the first. Do the same for Table D.

6. How does multiplying the y-coordinates of a figure by −1 change the location of a figure?

____

____

Table C

x	y
1	2
2	4
3	3
4	1

Table D

x	−y

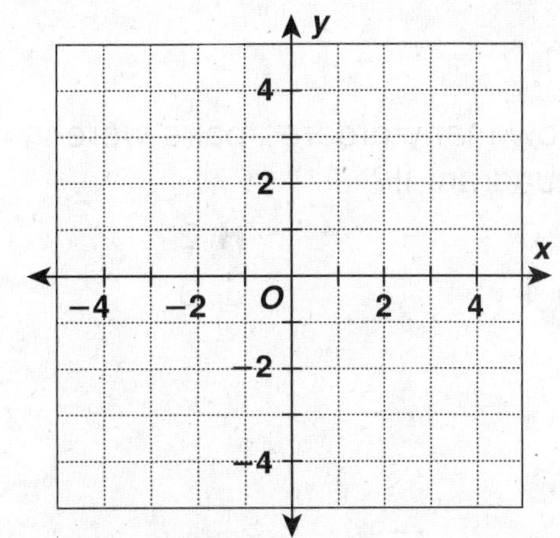

Holt Mathematics

LESSON 4-2

Problem Solving
Tables and Graphs

Write the correct answer.

1. The table shows the total cost of buying different numbers of bags of peanuts. Graph the data.

Number of Bags	1	2	3	4
Total Cost	$0.50	$1.00	$1.50	$2.00

2. What appears to be the relationship between the number of bags of peanuts and total cost shown in Problem 1?

____

____

3. How could you use the graph you made in Problem 1 to predict the total cost of 6 bags of peanuts?

____

____

4. Use the plan you described in Problem 3 to predict the total cost of 6 bags of peanuts.

____

Choose the letter for the best answer.

5. Terry plotted the ordered pairs from the table. How many ordered pairs were in Quadrant I?

A 0   **C** 2

B 1   **D** 3

x	y
−4	5
−1	3
2	1
5	−1

6. How many ordered pairs were in Quadrant II?

F 0   **H** 2

G 1   **J** 3

7. Which ordered pair is in Quadrant IV?

A (−4, 5)   **C** (2, 1)

B (−1, 3)   **D** (5, −1)

Holt Mathematics

Reading Strategies
Multiple Representations

You can represent pairs of x-values and y-values in three different ways.

Way 1: Make a **table of values**.

x	y
−5	−1
−2	1
1	3
4	5

Way 2: Write **ordered pairs**. Use the form (x, y).

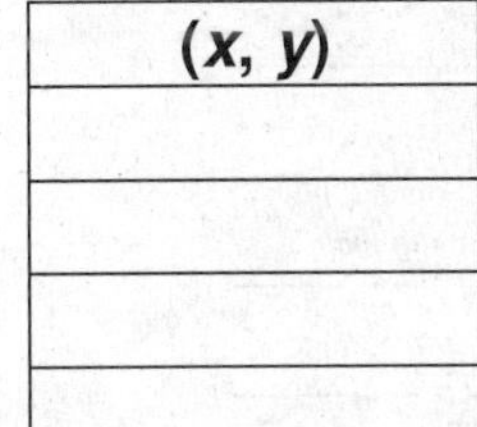

Way 3: Graph ordered pairs on a coordinate plane. Start at the origin. Move left or right for the x-value. Move up or down for the y-value

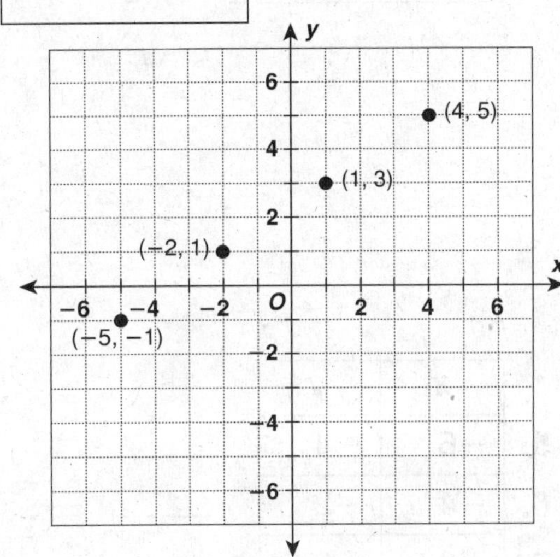

1. In the ordered pair (3, 5), which number is the x-value? ___________________

2. In the ordered pair (4, −8), which value is the y-value? ___________________

3. How can you use a graph to represent the ordered pair (−2, 5)?

____

4. Write an ordered pair for each pair of x- and y-values in the table. Then graph each ordered pair.

x	y
−1	−3
0	−1
1	1
2	3

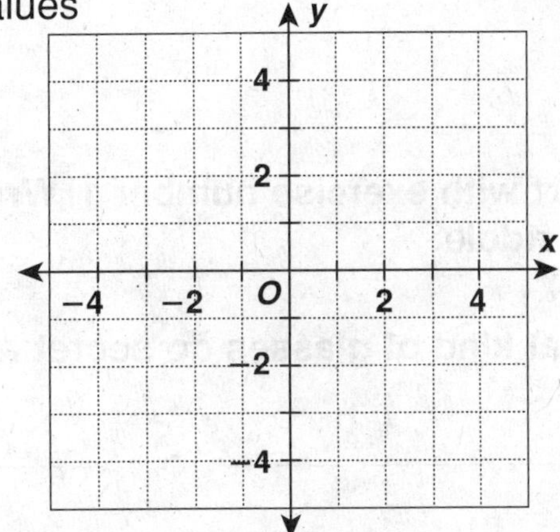

Holt Mathematics

Puzzles, Twisters & Teasers

LESSON 4-2

I Spy!

Find the point for each ordered pair on the coordinate plane.
Write the letter that names each point.

	x	y	
1.	−5	4	_______
2.	−3	1	_______
3.	−1	−2	_______
4.	1	−5	_______

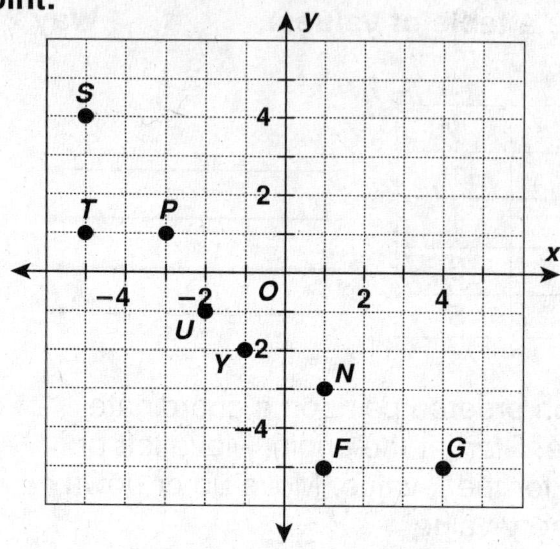

	x	y	
5.	−6	−4	_______
6.	−4	−2	_______
7.	−2	0	_______
8.	0	2	_______
9.	2	4	_______

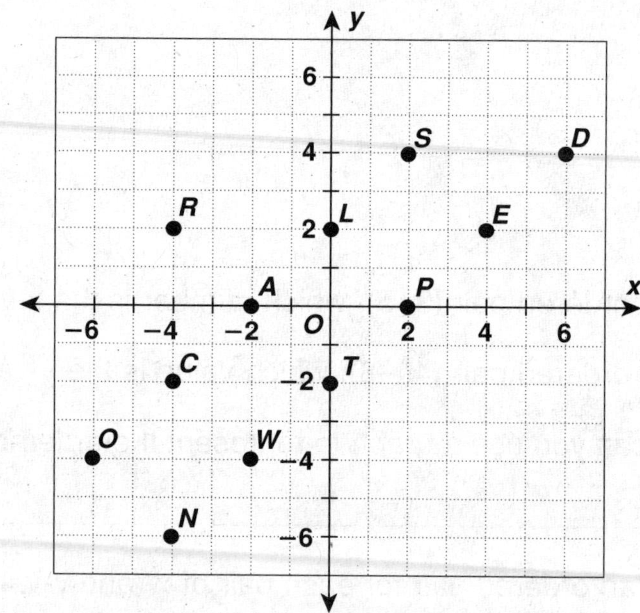

Start with exercise number 1. Write the letters in order to solve the riddle.

What kind of glasses do secret agents wear?

___ ___ ___ - ___ ___ ___ ___

18

Holt Mathematics

<table><tr><td>**LESSON**
4-3</td><td># Practice A
Interpreting Graphs</td></tr></table>

Match each graph with the description of the relationship it shows.

1. heating water to the boiling point

Graph ______

2. cooling water to the freezing point

Graph ______

3. pulling the plug in the bathtub

Graph ______

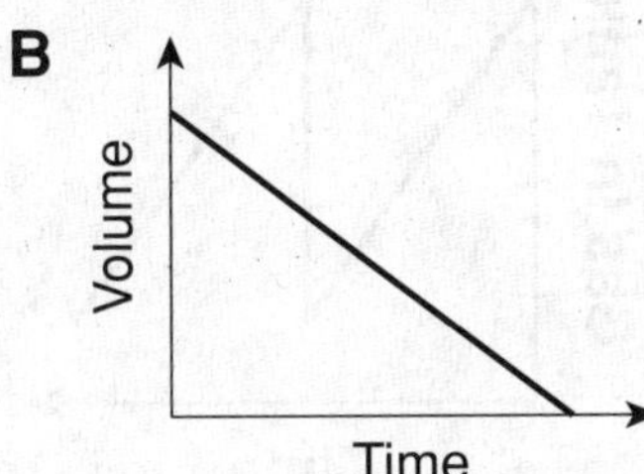

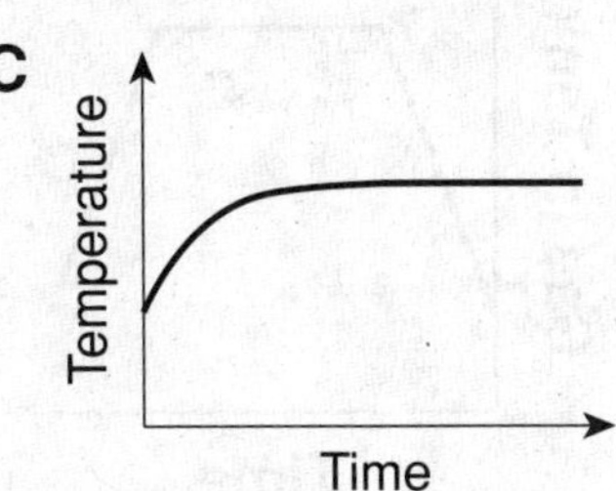

4. Ben withdrew money from his bank account each week for 6 weeks. Then he deposited money into the account. Which graph best shows the story? Circle the letter of your answer.

5. Talia hiked for 4 miles, and stopped to eat lunch. Then she hiked back to where she started. Complete the graph. Show the total distance Talia hiked compared to the time that she took.

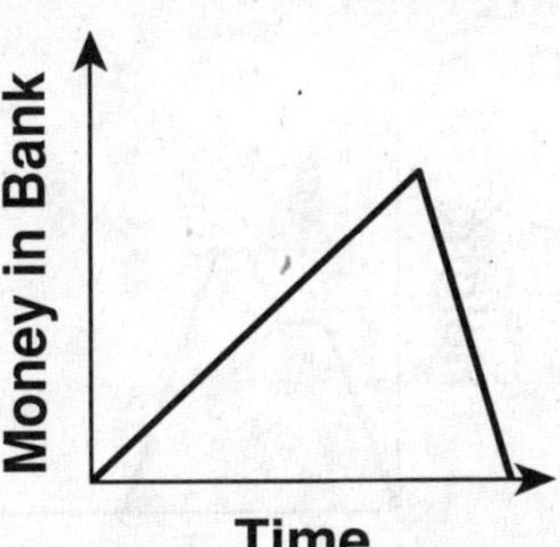

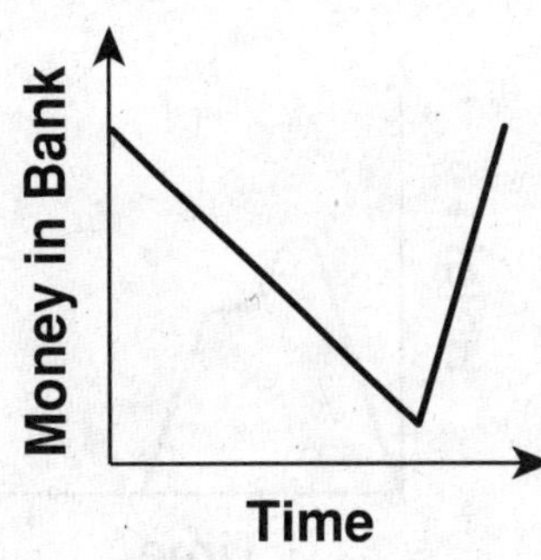

6. Use your graph to find the total number of miles Talia hiked.

____

Holt Mathematics

Practice B
Interpreting Graphs

1. The gas tank in Karen's car was full. Karen drove the car until only $\frac{1}{4}$ of the tank was full. Karen filled up the tank again and drove the car until $\frac{1}{4}$ of the tank was full. Which graph best shows the story? Circle the letter of your answer.

A

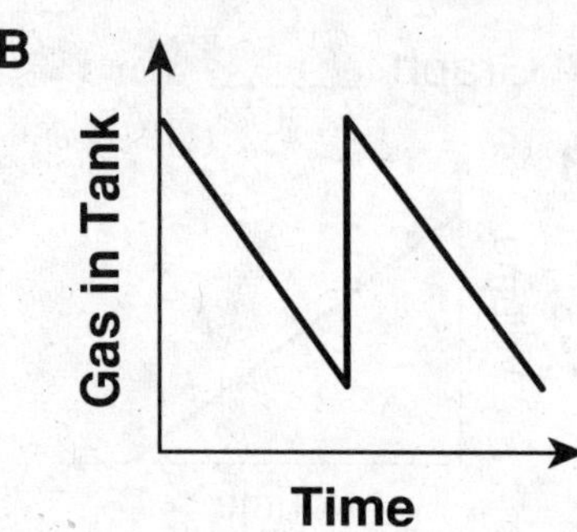

B

C

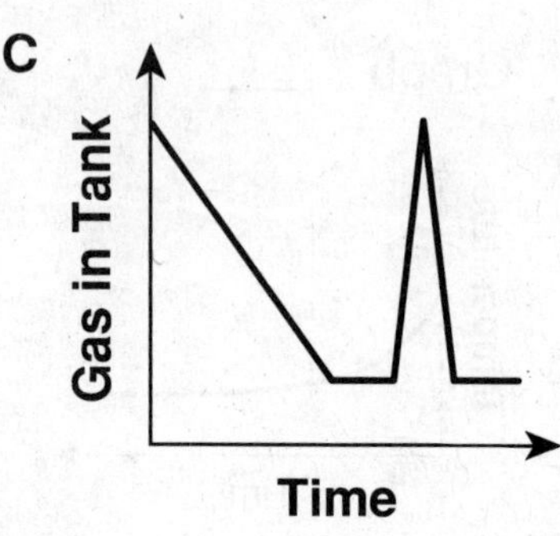

2. An elevator started at the ground floor. It went up to the sixth floor and stopped, then went to the fourth floor and stopped, and finally returned to the ground floor. Which graph best shows the story? Circle the letter of your answer.

F

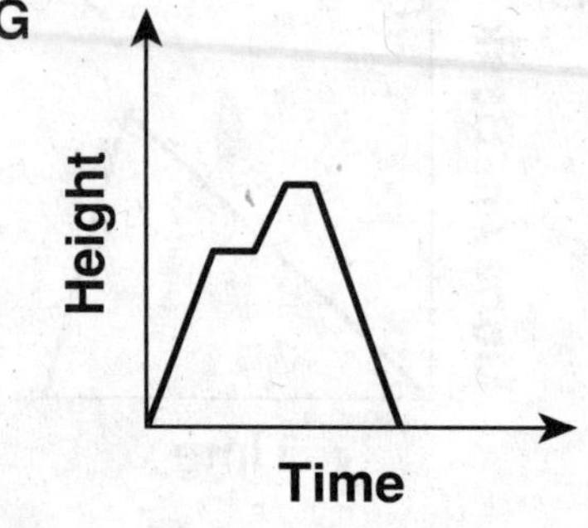

G

H

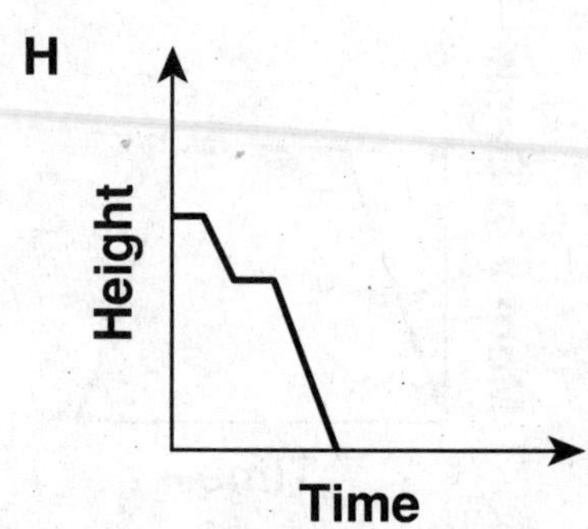

3. Maxine biked 6 miles from her house to the park. She played some softball. Then she biked 4 miles farther to the movie theater. After watching a movie, Maxine returned home. Complete the graph so that it shows the distance Maxine is from home compared to the time.

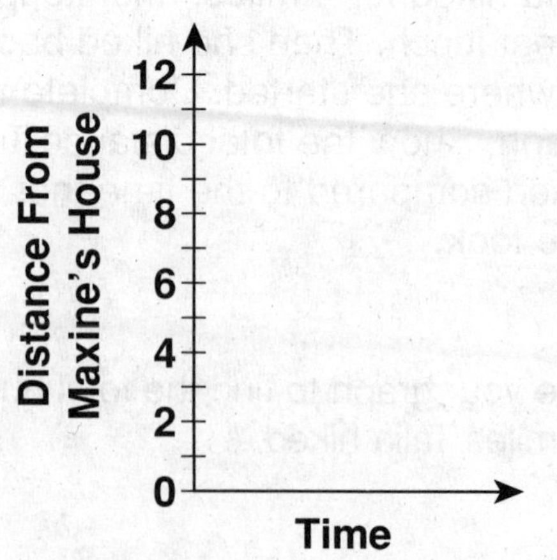

4. Use your graph to find the total number of miles Maxine biked.

Holt Mathematics

LESSON 4-3 | Practice C
Interpreting Graphs

1. An elevator was on the tenth floor of an office building. It came down and stopped at the fourth floor. It continued down, stopping at each of the remaining floors until it reached the first floor. Which graph best shows the story? Circle the letter of your answer.

A

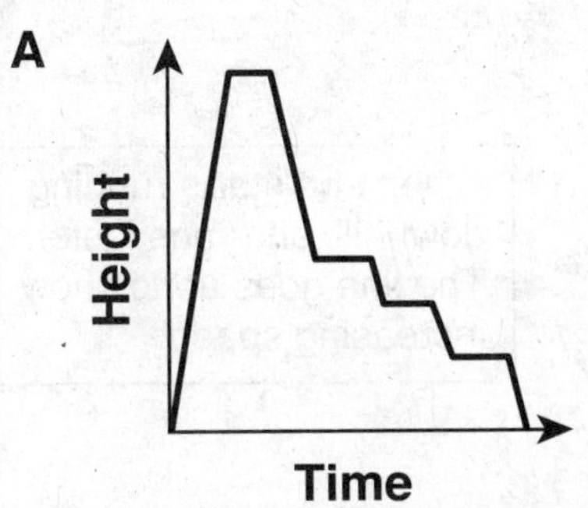

B

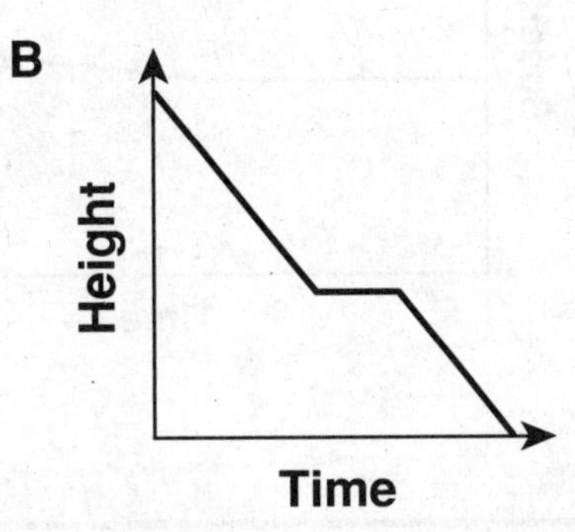

C

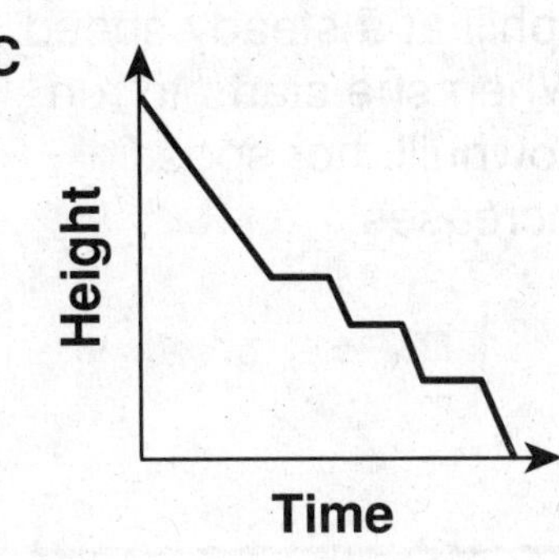

2. Yolanda took 16 weeks to save money for a trip. For the first 8 weeks, she saved $50 each week. For the next 4 weeks Yolanda did not save any money. Then Yolanda saved $50 a week for 4 weeks. When Yolanda took the trip, she spent $500 of her savings. Complete the graph so that it shows Yolanda's savings compared to time.

3. Use your graph to find the amount of money Yolanda had left after the trip.

____

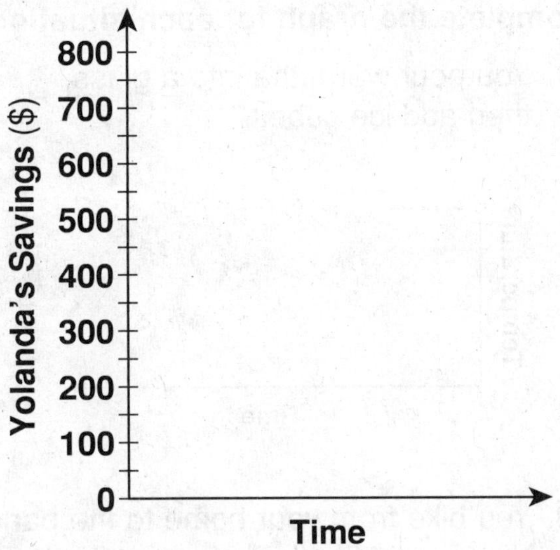

Write a story that describes the graph below.

4. The location of a skier on a ski slope ______________________________

____

____

____

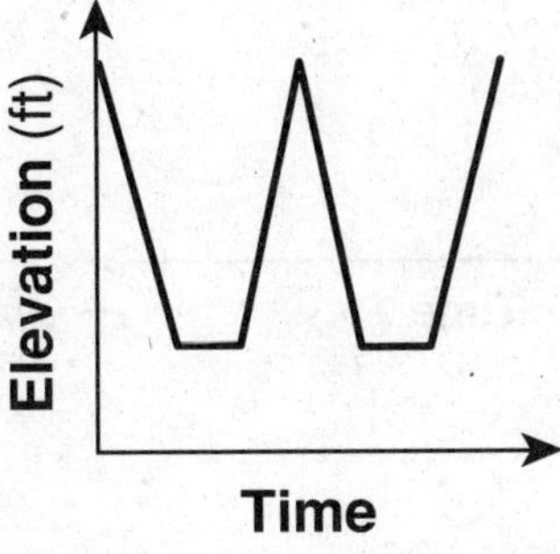

21

Holt Mathematics

Reteach
Interpreting Graphs

Graphs are often used to model situations. This graph shows Shavawn's daily jogging routine. She jogs uphill at a steady speed. When she starts to run downhill, her speed increases.

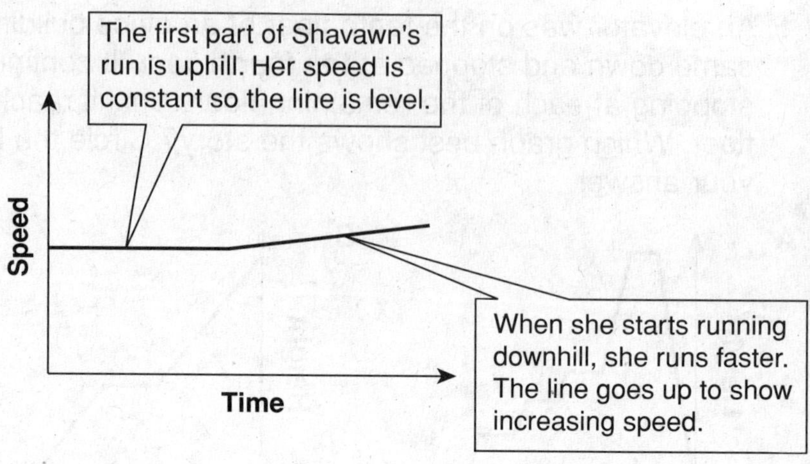

Complete the graph for each situation.

1. You pour warm tea into a glass, then add ice cubes.

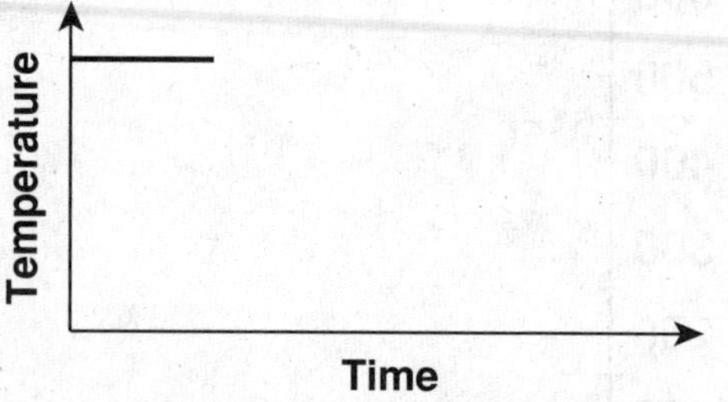

2. You are watching TV and lower the volume during a commercial.

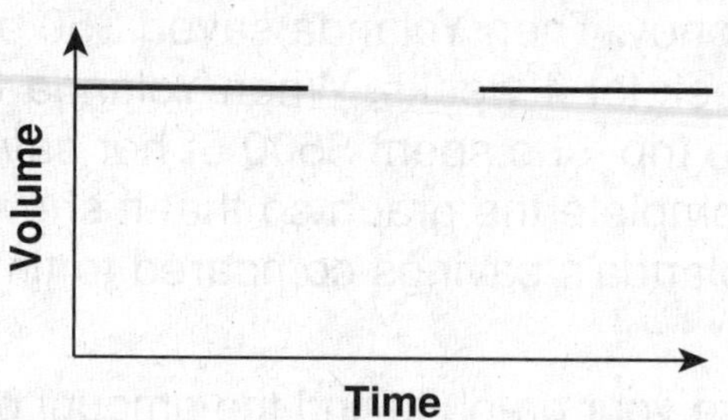

3. You bike from your home to the park. You play softball. Then you bike home.

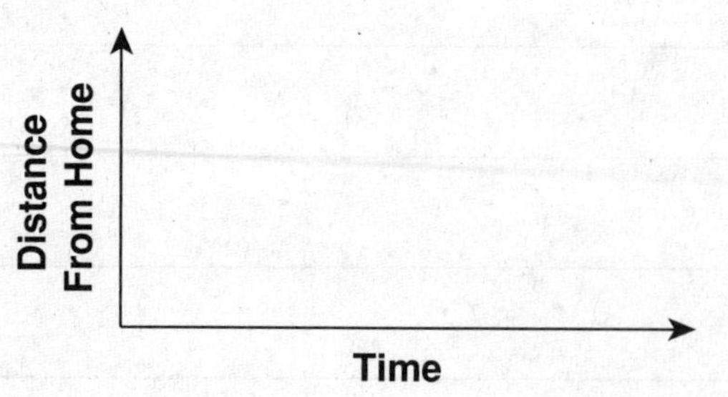

4. You drive at a steady speed. You hit a traffic jam, slow down, and travel at a slower speed. Then the traffic jam ends, and you travel at a faster speed.

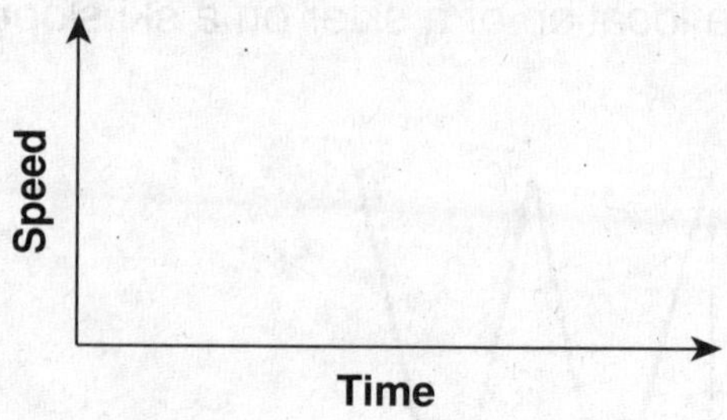

Holt Mathematics

LESSON 4-3 Challenge
Step Functions

A step graph is a graph that looks like steps. You can use step graphs to represent certain kinds of relationships. The step graph below shows the total cost of calls of different lengths on a particular pay phone system.

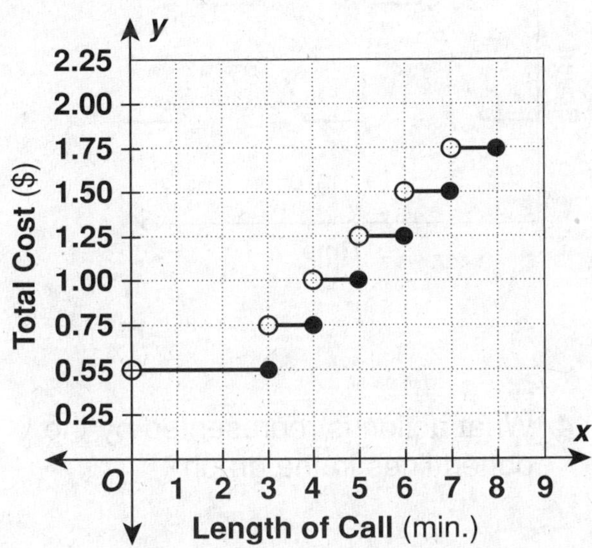

The symbol ○—● shows that the segment includes the right endpoint, but not the left endpoint.

Use the graph above to answer the questions.

1. What is the cost of making a 4-minute phone call? ___________________

2. What is the cost of making a 5.5-minute phone call? ___________________

3. Describe how the length of the call is used to determine the total cost.

____

____

4. Find the charge for any phone call with a length (l) that has the following values: 6 minutes < l ≤ 7 minutes. ___________________

5. How does this phone system calculate charges for fractions of minutes?

____

____

____

23

Holt Mathematics

<table><tr><td>**LESSON**
4-3</td><td></td></tr></table>

Problem Solving
Interpreting Graphs

Write the correct answer.

Eduardo exercises by doing laps around a track. He sprints the straight-aways and walks the curves.

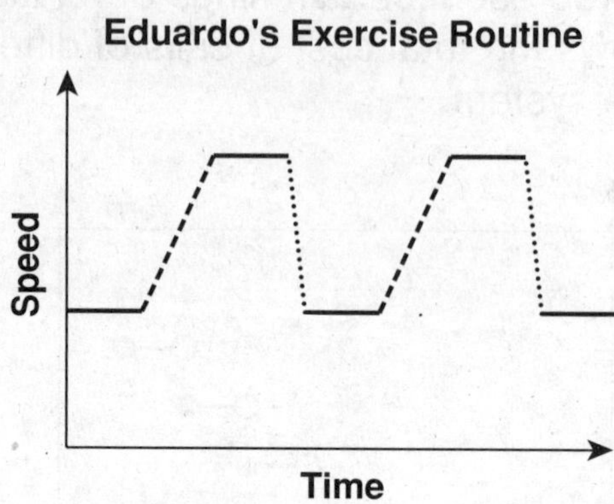

1. What action is represented by the lower horizontal lines in the graph?

   ____

2. What action is represented by the higher horizontal lines in the graph?

   ____

   ____

3. What action is represented by the dashed lines in the graph?

   ____

4. What action is represented by the dotted lines in the graph?

   ____

Choose the letter of the best answer.

5. What situation could this graph represent?

   A cost of birdseed by the pound

   B weight of bags of birdseed

   C ages of birds eating birdseed

   D a buy-one-get-one-free sale

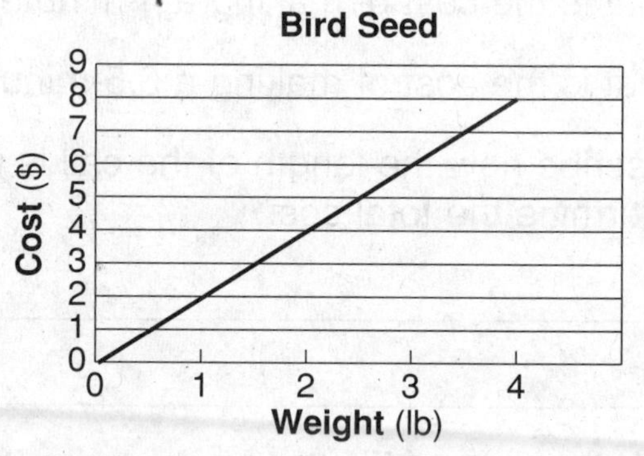

6. What would you expect to pay if you needed $3\frac{1}{2}$ pounds of birdseed?

   F $3.50

   G $5.00

   H $7.00

   J $7.50

7. What is the price for each pound of birdseed?

   A $0.25

   B $0.50

   C $1.00

   D $2.00

Holt Mathematics

LESSON 4-3 Reading Strategies
Read a Graph

Graphs are a good way to show that events may or may not relate to each other.

Graph A

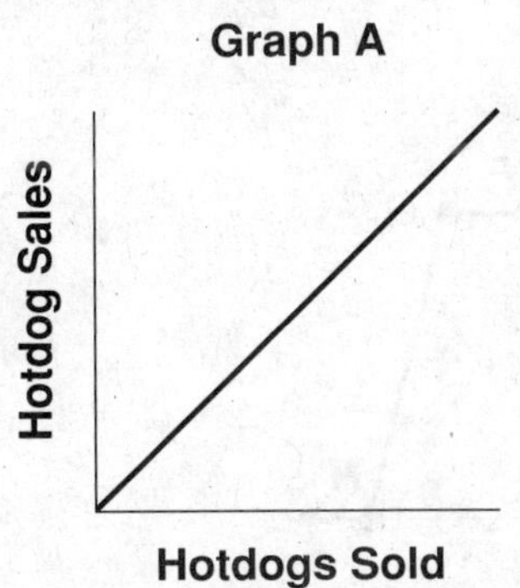

Graph B

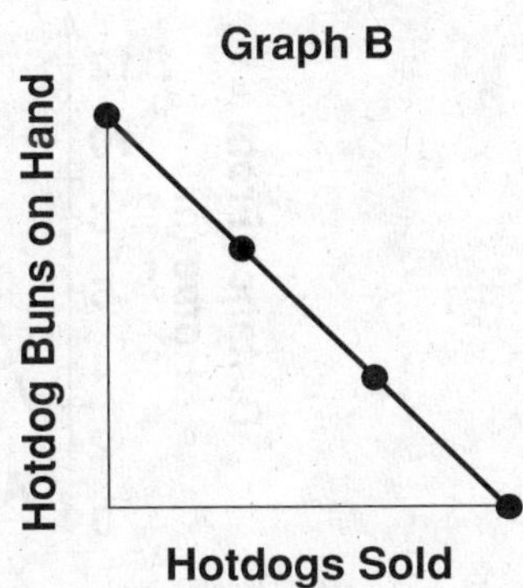

Use the graphs to answer each question.

1. Look at Graph A. As the number of hotdogs sold increases, what happens to hotdog sales?

____

2. Describe the direction of the line in Graph A.

____

3. Look at Graph B. As more hotdogs are sold, what happens to the number of hotdog buns on hand?

____

This graph compares the height of a hotdog stand with the number of hours the hotdog stand is open. Use the graph to answer each question.

4. Why is Graph C a horizontal line?

____

____

5. Describe the relationship between the height of the hotdog stand and the hours the stand is open.

____

Holt Mathematics

LESSON 4-3 Puzzles, Twisters, & Teasers
Graphically Speaking!

Use the graph to answer each question.
Circle the letter next to your answer.

Jenny's Distance from Home

1. How does the distance from home change between points C and D?
   E It increases.   **S** It decreases.   **R** It stays the same.

2. Which pair of points are the endpoints of a segment that represents an interval of time in which Jenny's distance from home does not change?
   F Points A and B   **I** Points D and E   **N** Points E and F

3. At point B, Jenny is at a flower store. How many miles from Jenny's home is the flower store?
   M 10 miles   **R** 8 miles   **S** 6 miles

4. When Jenny is furthest from home, how far from home is she?
   A 8 miles   **N** 10 miles   **F** 12 miles

5. Which of the following could describe what Jenny is doing between points G and H?
   T Jenny is traveling 8 miles to reach a store.

   O Jenny is stopped at a store that is 8 miles from home.

   L Jenny is traveling 8 miles to return home.

6. Suppose Jenny's home and the points at which she stops along the way are all along the same road. What is the total distance that Jenny travels?
   W 18 miles   **C** 20 miles   **G** 28 miles

Write the circled letters above the problem numbers to solve the riddle.

What is broken once you've spoken?

___ ___ ___ ___ ___ ___ ___
 3   2   5   1   4   6   1

26

Holt Mathematics

LESSON 4-4
Practice A
Functions, Tables, and Graphs

Find the output for each input.

1. $y = x + 5$

Input	Rule	Output
x	$x + 5$	y
0	$0 + 5$	
1	$1 + 5$	
2	$2 + 5$	

2. $y = 3x$

Input	Rule	Output
x	$3x$	y
0	$3(0)$	
1	$3(1)$	
2	$3(2)$	

3. $y = 3x - 4$

Input	Rule	Output
x	$3x - 4$	y
0	$3(0) - 4$	
1	$3(1) - 4$	
2	$3(2) - 4$	

4. $y = x^2$

Input	Rule	Output
x	x^2	y
0	$(0)^2$	
1	$(1)^2$	
2	$(2)^2$	

Make a function table. Graph the resulting ordered pairs.

5. $y = 2x - 4$

Input	Rule	Output	Ordered Pair
x	$2x - 4$	y	(x, y)
0	$2(0) - 4$		
1	$2(1) - 4$		
2	$2(2) - 4$		

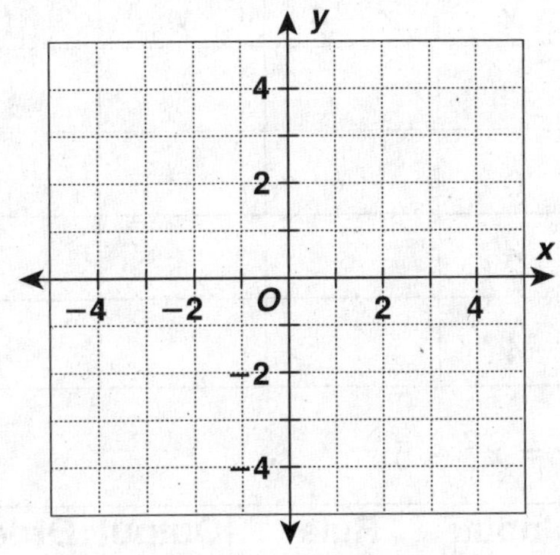

6. $y = 4x^2$

Input	Rule	Output	Ordered Pair
x	$4x^2$	y	(x, y)
-1	$4(-1)^2$		
0	$4(0)^2$		
1	$4(1)^2$		

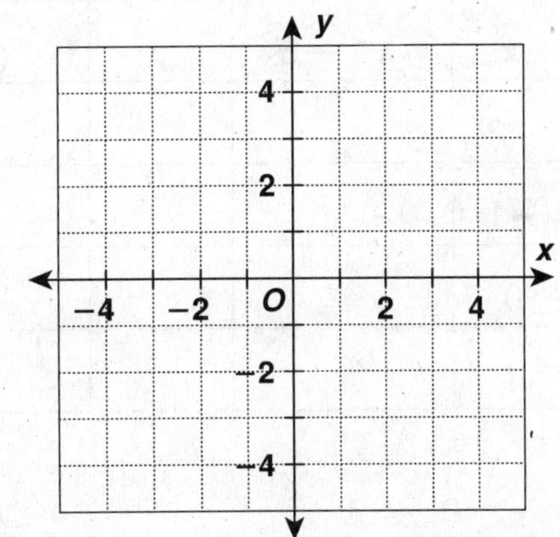

Holt Mathematics

Practice B
Functions, Tables, and Graphs

Find the output for each input.

1. $y = 5x - 1$

Input	Rule	Output
x	**5x – 1**	**y**
−2		
0		
3		
6		

2. $y = -2x^2$

Input	Rule	Output
x	**−2x²**	**y**
−2		
2		
3		
4		

Make a function table, and graph the resulting ordered pairs.

3. $y = x \div 4$

Input	Rule	Output	Ordered Pair
x	**x ÷ 4**	**y**	**(x, y)**
−4			
0			
2			
4			

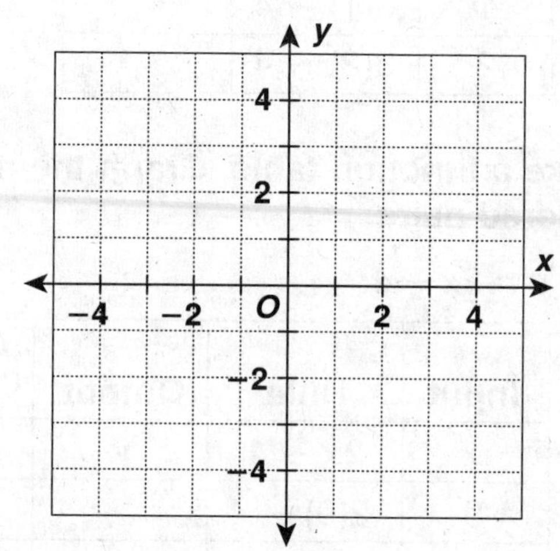

4. $y = x^2 - 5$

Input	Rule	Output	Ordered Pair
x	**x² – 5**	**y**	**(x, y)**
−2			
−1			
0			
1			

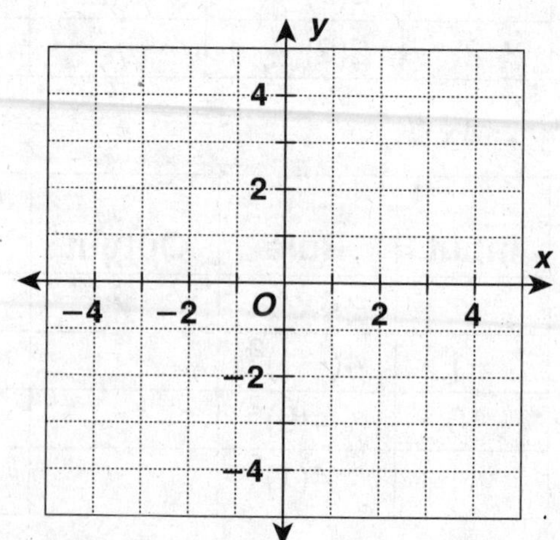

Holt Mathematics

Practice C

LESSON 4-4

Functions, Tables, and Graphs

Make a function table, and graph the resulting ordered pairs.

1. $y = \dfrac{x}{3}$

Input	Rule	Output	Ordered Pair
x	$\dfrac{x}{3}$	y	(x, y)
−3			
0			
3			

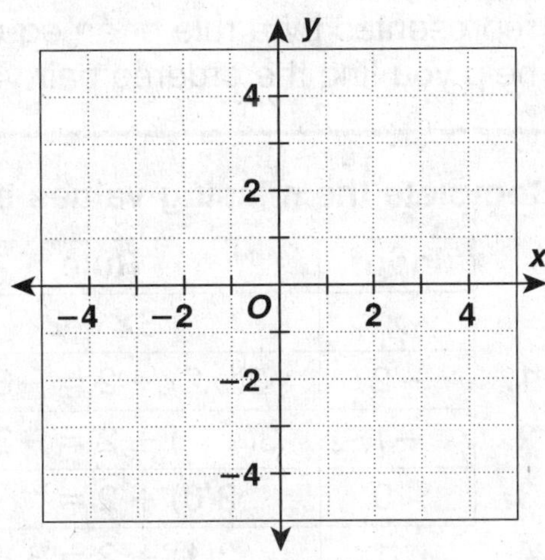

2. $y = 2x^2 - 4$

Input	Rule	Output	Ordered Pair
x	$2x^2 - 4$	y	(x, y)
−2			
0			
1			
2			

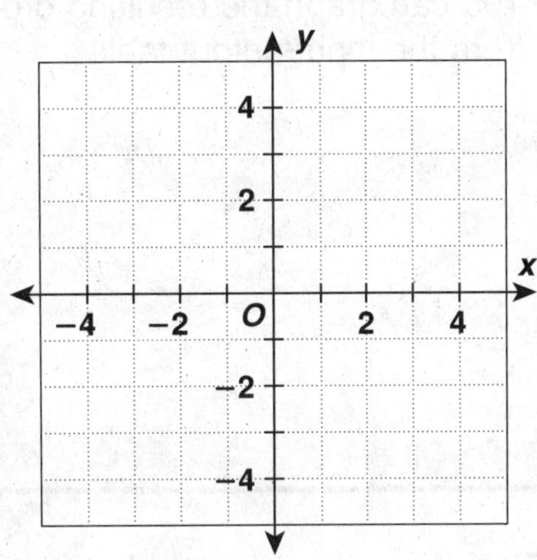

3. For every dollar you spend at the Rusty Nail hardware chain, you receive 20 Gold Hammer points that can be redeemed for free merchandise. The equation $y = 20x$ gives the number of points y received for spending x dollars. Make an input/output table using the values $x = 10, 25, 50, 80$.

Input	Rule	Output
x		y

Holt Mathematics

LESSON 4-4 Reteach
Functions, Tables, and Graphs

A **function** is a relationship in which the value of one quantity depends on the value of another quantity. A function can be represented by a rule or an equation. A *function table* can help you find the ordered pair values for a function.

Complete the missing values in the table.

	Input	Rule	Output	Ordered Pair
	x	**3x + 2**	**y**	**(x, y)**
1.	−2	3(−2) + 2 = −6 + 2		(−2, −4)
2.	−1	3( ) + 2 = −3 + 2	−1	(, −1)
3.	0	3(0) + 2 = + 2	2	(, 2)
4.	1	3(1) + 2 = 3 + 2		(,)

You can graph the resulting ordered pairs from the input/output table.

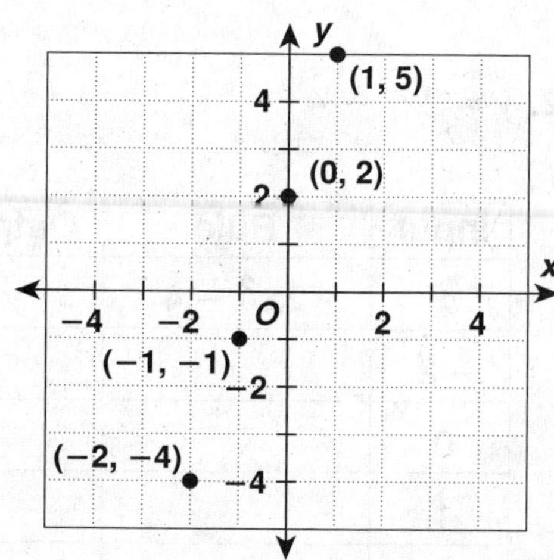

Find the output for each input value, and graph the resulting ordered pairs.

5. $y = 2x - 1$

Input	Rule	Output	Ordered Pair
x	**2x − 1**	**y**	**(x, y)**
−2			
−1			
0			
1			
2			

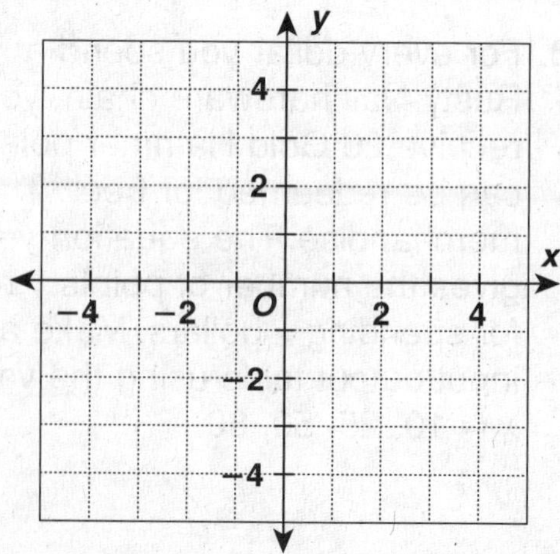

Holt Mathematics

LESSON 4-4 Challenge
Function Notation

You can use the equation of a function to find values for *y* that satisfy the equation for any value of *x*.

For example, if $y = 3x + 1$, then for input $x = 2$, the output is $y = 7$.

Function notation allows you to keep track of both input and output. In function notation, the variable *y* is written as $f(x)$. This reminds you that the value of the function, $f(x)$, depends on the value of *x*.

Suppose $f(x) = 3x + 1$.

To find $f(2)$, substitute 2 for *x* in the equation.

$$f(2) = 3(2) + 1$$
$$f(2) = 7$$

Find $g(-3)$ if $g(x) = x^2 + 1$

$$g(-3) = (-3)^2 + 1 = 9 + 1$$
$$g(-3) = 10$$

Find $h(7)$ if $h(x) = (x - 5)^2$

$$h(7) = (7 - 5)^2 = (2)^2$$
$$h(7) = 4$$

Evaluate using function notation.

1. Let $f(x) = 3x + 1$. Find $f(0)$, $f(-1)$, and $f(1)$.

2. Let $g(x) = x^2 + 1$. Find $g(0)$, $g(1)$, and $g(5)$.

3. Let $f(x) = (x - 5)^2$. Find $f(0)$, $f(1)$, and $f(-1)$.

4. Let $h(x) = 2x^2$. Find $h(0)$, $h(-1)$, and $h(2)$.

5. Let $f(x) = 8x - 11$. Find $f(0)$, $f(-3)$, and $f(5)$.

6. Let $g(x) = x^2 - 1$. Find $g(0)$, $g(1)$, and $g(-4)$.

7. Let $h(x) = (x + 1)^2$. Find $h(0)$, $h(1)$, and $h(-1)$.

Holt Mathematics

Problem Solving
Functions, Tables, and Graphs

Write the correct answer.

1. Film passes through a projector at the rate of 24 frames per second. The equation $y = 24x$ describes the number of frames, y, that have passed over any number of seconds, x. Complete the function table.

Input	Rule	Output
x		y
2		48
3		72
4		96
5		120
6		144

2. Anne pays $40 a month for cable TV, plus $4 for each movie she watches on pay-per-view channels. Complete the function table, where x is the number of pay-per-view movies she watches each month and y is her monthly cable bill.

Input	Rule	Output
x		y
3		52
5		60
7		68
9		76
11		84

Choose the letter of the best answer.

Madeline has a discount coupon for $5 off her next purchase of tennis balls. Tennis balls are on sale for $3 per can. The equation $y = 3x - 5$ gives her final cost, y, to purchase x cans of tennis balls.

3. If $x = 3$, what is the value of y?
   A 15
   B 14
   C 9
   D 4

4. If $x = 5$, what is the value of y?
   F 25
   G 15
   H 10
   J 5

5. If $x = 10$, what is the value of y?
   A 25
   B 30
   C 35
   D 50

6. If $x = 4$, what is the value of y?
   F 5
   G 7
   H 12
   J 17

7. If $x = 9$, what is the ordered pair (x, y)?
   A (5, 9)
   B (9, 5)
   C (9, 22)
   D (9, 10)

8. If $x = 6$, what is the ordered pair (x, y)?
   F (6, 18)
   G (6, 30)
   H (12, 31)
   J (6, 13)

Holt Mathematics

Reading Strategies
4-4 *Focus On Vocabulary*

A **function** is a special rule that pairs exactly one output value to each input value.

The value that replaces a variable is called an **input** value. The resulting value is called the **output**.

This equation shows the location of input and output in this function.

output   **input**

$$y = 3x + 4$$

This table lists input and output values for $y = 3x + 4$.

Input	Rule	Output
x	$3x + 4$	y
1	$3(1) + 4$	7
2	$3(2) + 4$	10
3	$3(3) + 4$	13
4	$3(4) + 4$	16

Answer the following questions.

1. Which variable stands for the input value?

   ____

2. Which variable stands for the output value?

   ____

3. What do you call the special relation or rule that exists between the input value and the output value?

   ____

4. An input value in a function can be paired with how many output values?

   ____

5. In the table, which input value is paired with the output value 10?

   ____

Holt Mathematics

Puzzles, Twisters & Teasers
Get to the Bottom of It!

Fill in the blanks to complete the function table.

Problem #	Input	Rule	Output	Ordered Pair
1	x	$2x$	y	(x,y)
2	−2	2(−2)		(−2,−4)
3	−1	2(−1)	−2	
4	0		0	(0,0)
5	1	2(1)		(1,2)
6	2	2(2)	4	
7	3	2(3)		
8	4		8	(4,8)
9	5	2(5)		
10	6	2(6)		(6,12)

You've dropped your locker key in the pool and you must go through the maze to find it.

Step 1 – Go to start

Step 2 – Go right the number of the output in problem #9

Step 3 – Go down the number of the output in problem #7

Step 4 – Go left the absolute value of the output in problem #2

Step 5 – Go up the *y* value of the ordered pair in problem #6

Step 6 – Go right the *x* value of the ordered pair in problem #9

Sept 7 – Go down the *y* value in the ordered pair in problem #7

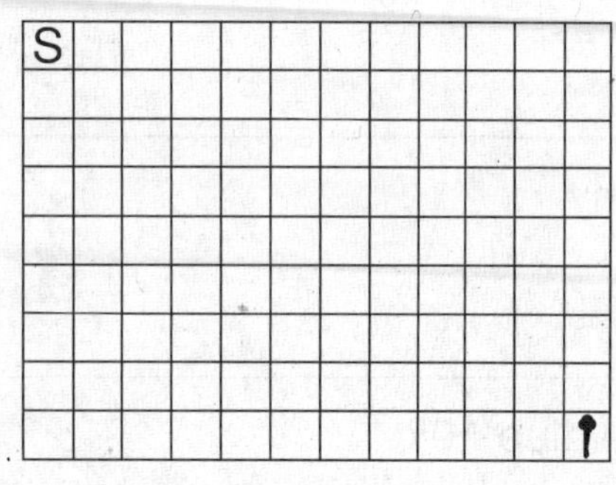

Holt Mathematics

Practice A

Find a Pattern in Sequences

Tell whether each sequence of *y*-values is arithmetic or geometric. Then find *y* when $n = 5$.

1.

n	1	2	3	4	5
y	3	6	12	24	■

____

2.

n	1	2	3	4	5
y	7	12	17	22	■

____

3.

n	1	2	3	4	5
y	12	30	48	66	■

____

4.

n	1	2	3	4	5
y	4	12	36	108	■

____

Choose the function from the box that describes each sequence.

$y = 4n$	$y = n - 4$	$y = n + 4$	$y = n - 1$	$y = n - 9$
$y = n + 6$	$y = n + 8$	$y = 8n$	$y = n - 0.1$	$y = 9n$

5. −3, −2, −1, 0, ...

____

6. 8, 16, 24, 32, ...

____

7. 9, 10, 11, 12, ...

____

8. 0.9, 1.9, 2.9, 3.9, ...

____

9. 4, 8, 12, 16, ...

____

10. 7, 8, 9, 10, ...

____

11. 0, 1, 2, 3, ...

____

12. 5, 6, 7, 8, ...

____

13. 9, 18, 27, 36, ...

____

14. −8, −7, −6, −5, ...

____

Holt Mathematics

Practice B
Find a Pattern in Sequences

Tell whether each sequence of *y*-values is arithmetic or geometric. Then find *y* when $n = 5$.

1.

n	1	2	3	4	5
y	−5	10	25	40	■

2.

n	1	2	3	4	5
y	14	28	56	112	■

3.

n	1	2	3	4	5
y	4	12	36	108	■

4.

n	1	2	3	4	5
y	28	42	56	70	■

Write a function that describes each sequence.

5. 12, 24, 36, 48, ...

6. 13, 14, 15, 16, ...

7. −8, −7, −6, −5, ...

8. 2.5, 3.5, 4.5, 5.5, ...

9. 9, 18, 27, 36, ...

10. −12, −11, −10, −9, ...

11. $\frac{3}{4}, 1\frac{3}{4}, 2\frac{3}{4}, 3\frac{3}{4}, ...$

12. −3, −6, −9, −12, ...

13. Mike ran 4 km on Sunday, 6 km on Monday, and 8 km on Tuesday. Write a function that describes the sequence. Then use the function to predict how many kilometers Mike will run on Friday.

Holt Mathematics

Practice C
Find a Pattern in Sequences

Tell whether each sequence of *y*-values is arithmetic or geometric. Then find *y* when *n* = 5.

1.

n	1	2	3	4	5
y	−3	9	−27	81	■

2.

n	1	2	3	4	5
y	5	3.5	2	0.5	■

____

Write a function that describes each sequence.

3. 7, 13, 19, 25, …

4. 0.8, 1.2, 1.6, 2.0, …

5. 1, 5, 9, 13, …

6. −20, −15, −10, −5, …

7. $\dfrac{4}{3}, \dfrac{5}{3}, \dfrac{6}{3}, \dfrac{7}{3}, \ldots$

8. −7, −5, −3, −1, …

Find a function that describes each sequence. Use the function to find the tenth term in the sequence.

9. −7, −3, 1, 5, …

10. 8, 17, 26, 35, …

11. 2.2, 3.2, 4.2, 5.2, …

12. −12, −6, 0, 6, …

13. Maria uses snap cubes to make a series of cubes. She makes the first cube by showing 1 snap cube. She makes the second cube by connecting 8 snap cubes, and the third cube by arranging 27 snap cubes. Write a function that describes this sequence. Then use the sequence to predict the number of snap cubes Maria will use to make the sixth cube.

____

Holt Mathematics

LESSON 4-5 Reteach
Find a Pattern in Sequences

Making a table is helpful for finding a function rule for a sequence of numbers.

Complete the tables for each sequence.

1. 4, 8, 12, 16, ...

Term n	1	2	3	4	5	6	7
Value y	4	8	12	16			

Rule: Add _______ to each term to find the value of the next term.

Notice that the value of each term is 4 times the term number.

Term n	1	2	3	4	5	6	7
Rule	4 • 1	4 • 2	4 • 3	4 • 4			
Value y	4	8	12	16	20	24	28

Write a function to describe this sequence.

Multiply the term number by _______, or $y = $ _______n.

2. 6, 7, 8, 9, ...

Term n	1	2	3	4	5	6	7
Value y	6	7	8	9			

Rule: Add _______ to each term to find the value of the next term.

Notice that the value of each term is 5 more than the term number.

Term n	1	2	3	4	5	6	7
Rule	1 + 5	2 + 5	3 + 5	4 + 5			
Value y	6	7	8	9			

Write a function to describe this sequence.

Add _______ to the term number, or $y = n + $ _______

Write a function that describes each sequence.

3. 6, 12, 18, 24, ...

____

4. 11, 12, 13, 14, ...

____

Holt Mathematics

Challenge
Find the Term

You can use the pattern in a sequence to write an expression for the *n*th term of the sequence. Then you can use this expression to find any term in the sequence.

Find the fiftieth term of the following sequence: $\frac{1}{1}, \frac{1}{2}, \frac{1}{3}, \frac{1}{4}, \ldots$

- The first term in this sequence is $\frac{1}{1}$. The numerator and the denominator of the fraction are both the same number as the location of the term in the sequence.

- The second term in this sequence is $\frac{1}{2}$. The numerator is 1, and the denominator is 2, the same number as the location of the term in the sequence.

- The third term in this sequence is $\frac{1}{3}$. The numerator is 1 and the denominator is 3, the same number as the location of the term in the sequence.

From the pattern, the *n*th term is $\frac{1}{n}$. So, the fiftieth term is $\frac{1}{50}$.

Solve.

1. Find the *n*th term and the twentieth term of the following sequence:
   −2, −1, 0, 1, …

2. Find the *n*th term and the fortieth term of the following sequence:
   9, 10, 11, 12, …

3. Find the *n*th term and the fifteenth term of the following sequence:
   5, 10, 15, 20, …

4. Find the *n*th term and the twenty-fifth term of the following sequence:
   −2, −4, −6, −8, …

5. Find the *n*th term and the twentieth term of the following sequence: $\frac{1}{2}, \frac{2}{3}, \frac{3}{4}, \frac{4}{5}, \ldots$

6. Find the *n*th term and the fiftieth term of the following sequence: $\frac{1}{2}, \frac{1}{4}, \frac{1}{6}, \frac{1}{8}, \ldots$

Holt Mathematics

Problem Solving
Find a Pattern in Sequences

Write the correct answer.

1. Marina earns $15 for 1 hour of babysitting, $23 for 2 hours of babysitting, and $31 for 3 hours of babysitting. Write a function to describe the sequence.

____

2. A website has 175 hits during its first hour of operation. It has 350 hits during its second hour and 525 hits during its third hour. Write a function to describe the sequence.

____

3. Rodney had 53 baseball cards. Then he started to buy cards each week. After 1 week, Rodney had 73 cards. After 2 weeks, he had 93 cards. After 3 weeks, he had 113 cards. Write a function to describe the sequence. Then use it to predict the number of cards Rodney will have after 5 weeks.

____

4. Jen is reading a novel that is 463 pages long. After the first day, she had 423 pages left. After 2 days, she had 383 pages left. After 3 days, she had 343 pages left. Write a function to describe the sequence, and use it to predict the number of pages that Jen will have left after 8 days.

____

Choose the letter of the best answer.

Use the sequence 3, 6, 9, 12, … for the following exercises.

5. What is the rule for this sequence?
   A Add 3 to n.
   B Subtract 3 from n.
   C Multiply n by 3.
   D Divide n by 3.

6. What is the function that describes this sequence?
   F $y = 3n$
   G $y = \dfrac{n}{3}$
   H $y = n + 3$
   J $y = n - 3$

7. What are the next 3 terms in the sequence?
   A 9, 6, 3
   B 15, 18, 21
   C 15, 19, 24
   D 36, 432, 5,184

8. What is the ninth term in the sequence?
   F 24
   G 27
   H 30
   J 33

Holt Mathematics

Reading Strategies
Analyze Information

A **sequence** is a list of values arranged in a special order or pattern. Numbering the terms in a sequence helps you analyze the pattern.

Term	1st	2nd	3rd	4th
Value	3	6	9	12

1. What do you call a list of numbers that create a pattern?

____

2. What is the value of the first term in the sequence above?

____

3. The number 9 is the value for which term in the sequence?

____

4. What pattern can you see for the values in the table?

____

A function table can help you see the relation between the number of the term and value of the term.

Number of term: n	Rule	Value of term: y
1	$1 \cdot 3$	3
2	$2 \cdot 3$	6
3	$3 \cdot 3$	9
4	$4 \cdot 3$	12

Study the function table to complete each question.

5. When the value of n is 2, what is the value of y?

____

6. When the value of n is 4, what is the value of y?

____

7. What pattern do you see for the value of y?

____

8. What would be the values of the fifth term and the sixth term? _______________

Puzzles, Twisters, & Teasers

LESSON 4-5 *Weigh the Options!*

Decide whether each sequence of y-values is arithmetic or geometric. Circle the letter above your answer.

1.

n	1	2	3	4
y	10	20	30	40

T   **A**
arithmetic   geometric

2.

n	1	2	3	4
y	7	21	63	189

F   **H**
arithmetic   geometric

3.

n	1	2	3	4
y	2	8	32	128

U   **I**
arithmetic   geometric

4.

n	1	2	3	4
y	36	30	24	18

R   **S**
arithmetic   geometric

Next, choose the function that describes each sequence. Circle the letter above your answer.

5. $-5, -4, -3, -2, \ldots$

    L   **W**

$y = n - 5$ $y = n - 6$

6. $3.5, 4.5, 5.5, 6.5, \ldots$

    D   **N**

$y = 3.5n$ $y = n + 2.5$

7. $30, 60, 90, 120, \ldots$

    C   **H**

$y = 30n$ $y = n + 30$

8. $\dfrac{1}{4}, 1\dfrac{1}{4}, 2\dfrac{1}{4}, \ldots$

    A   **O**

$y = n - \dfrac{3}{4}$ $y = n + \dfrac{1}{4}$

9. $-9, -8, -7, -6, \ldots$

    T   **E**

$y = -9n$ $y = n - 10$

10. $25, 50, 75, 100, \ldots$

    S   **D**

$y = 25n$ $y = n + 25$

Start with problem number 1. Write the circled letters above the problem numbers to solve the riddle.

Why are fish easy to weigh?

They have

___ ___ _E_ ___ ___ ___ _O_ ___ ___
1.   2.   3.   4.   5.   6.

___ _S_ ___ ___ _L_ ___ ___ .
   7.   8.   9.   10.

Holt Mathematics

Practice A

LESSON 4-6

Graphing Linear Functions

Complete the function tables. Then match the letter of each graph with the function table for its linear function.

1. $y = x - 2$ Graph: ______

Input	Linear Equation	Output	Ordered Pair
x	$y = x - 2$	y	(x, y)
0			
1			
2			

A

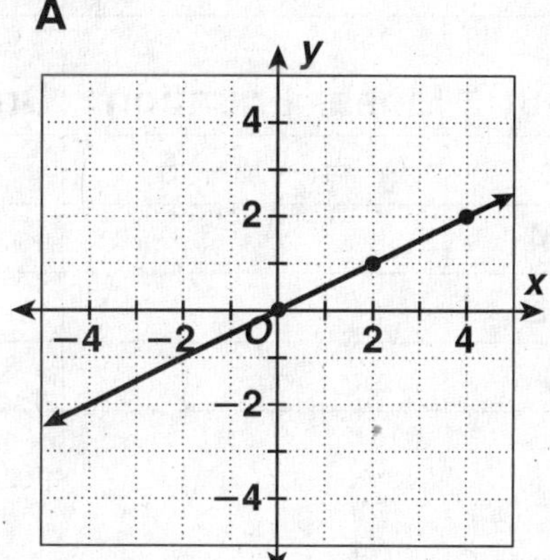

2. $y = \dfrac{x}{2}$ Graph: ______

Input	Linear Equation	Output	Ordered Pair
x	$y = \dfrac{x}{2}$	y	(x, y)
0			
2			
4			

B

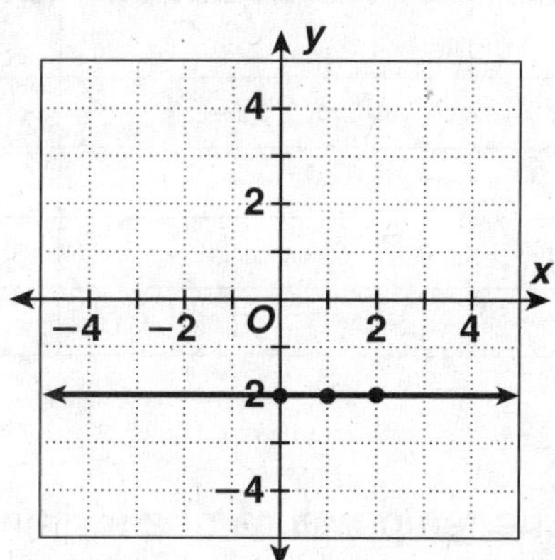

3. $y = -2$ Graph: ______

Input	Linear Equation	Output	Ordered Pair
x	$y = -2$	y	(x, y)
0			
1			
2			

C

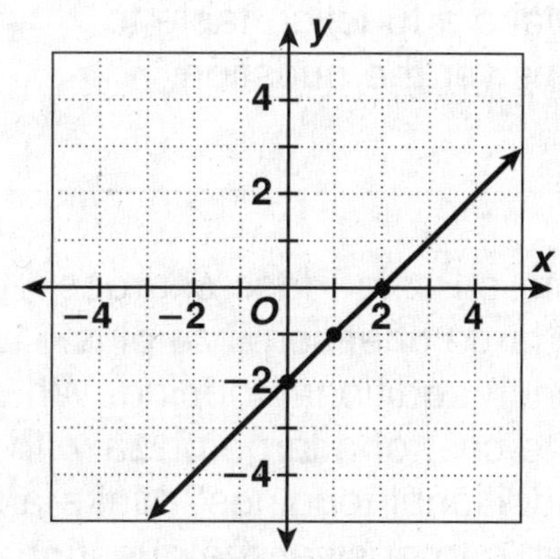

Holt Mathematics

Practice B
Graphing Linear Functions

Graph each linear function.

1. $y = -x - 5$

Input	Linear Equation	Output	Ordered Pair
x	$y = -x - 5$	**y**	**(x, y)**
−4			
−2			
0			

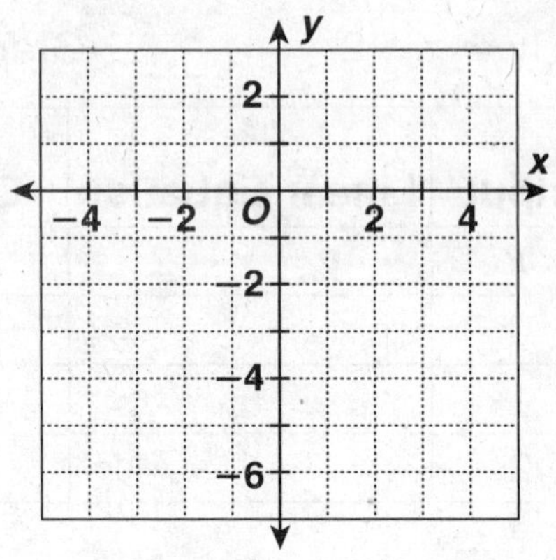

2. $y = 2x - 1$

Input	Linear Equation	Output	Ordered Pair
x	$y = 2x - 1$	**y**	**(x, y)**
−2			
0			
1			

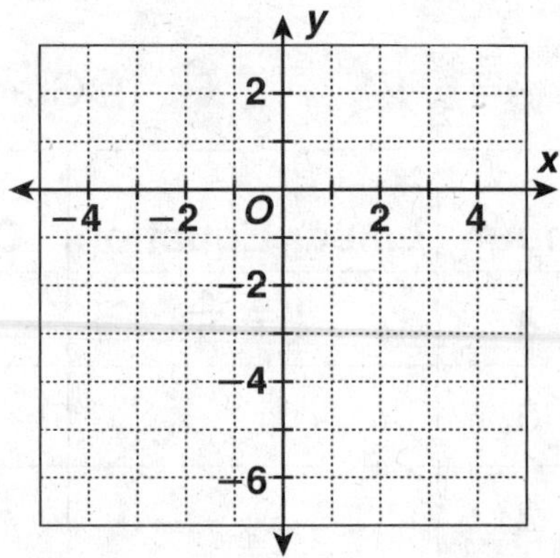

3. The temperature of a swimming pool is 75°F. When the pool heater is turned on, the temperature rises 2°F every hour. What will the temperature be after 3 hours? Make a function table to answer the question.

4. Mel's Pizza Place charges $15.00 for a large cheese pizza plus $1.25 for each additional topping. What will be the cost of a large pizza with 3 additional toppings? Make a function table to answer the question.

Holt Mathematics

LESSON 4-6 — Practice C
Graphing Linear Functions

Graph each linear function.

1. $y = x - 4$

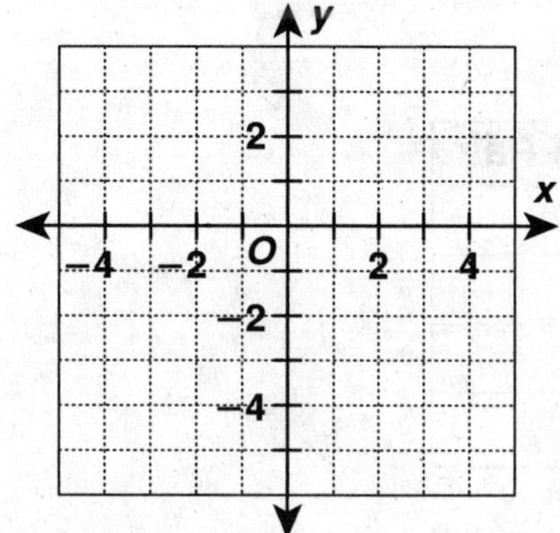

2. $y = \dfrac{x}{3} - 3$

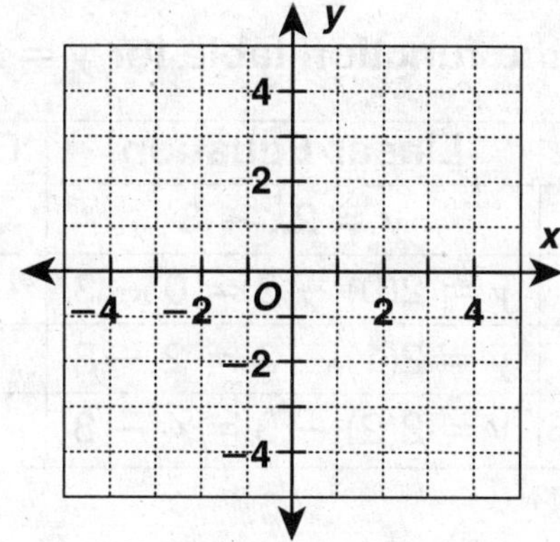

3. $y = x + 3$

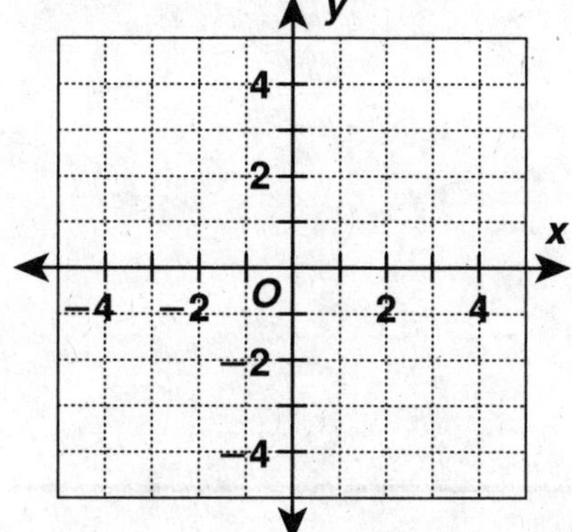

4. $y = 3x + 4$

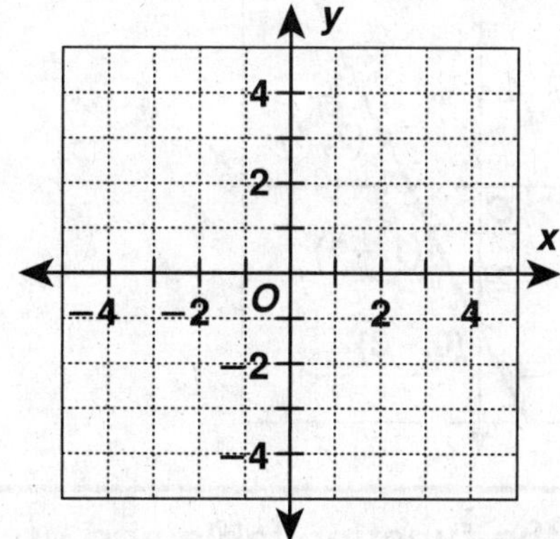

5. Jackie is saving her money to buy a coat that costs $121. If she already has $61 and saves $15 each week, in how many weeks can she buy the coat? Make a function table to answer the question.

____

Holt Mathematics

LESSON 4-6 Reteach
Graphing Linear Functions

The graph of a linear equation is a straight line. A **linear function** is
a function whose graph is a straight line that is not vertical. To graph
a linear equation, first make a function table.

Complete the function table for $y = 2x - 3$.

	Input	Linear Equation	Output	Ordered Pair
	x	**$y = 2x - 3$**	**y**	**(x, y)**
1.	0	$y = 2(0) - 3 = 0 - 3$		
2.	1	$y = 2(1) - 3 = 2 - 3$		
3.	2	$y = 2(2) - 3 = 4 - 3$		

To graph the equation, graph ordered pairs. Then draw a line through the points.

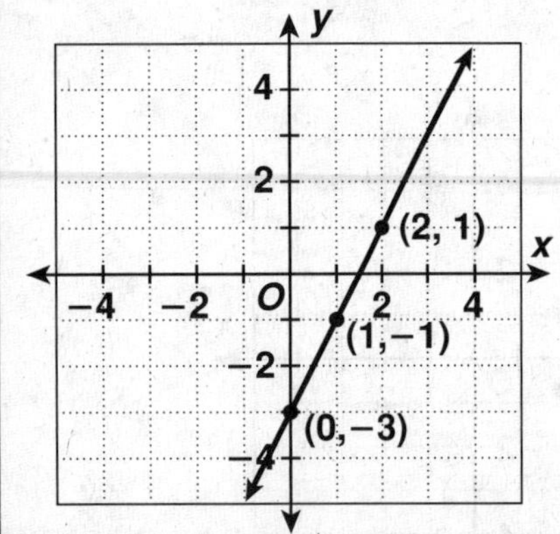

Complete the function table and graph the equation.

4. $y = -x + 1$

Input	Linear Equation	Output	Ordered Pair
x	**$y = -x + 1$**	**y**	**(x, y)**
0	$y = -0 + 1$		
1	$y = -1 + 1$		
2	$y = -2 + 1$		

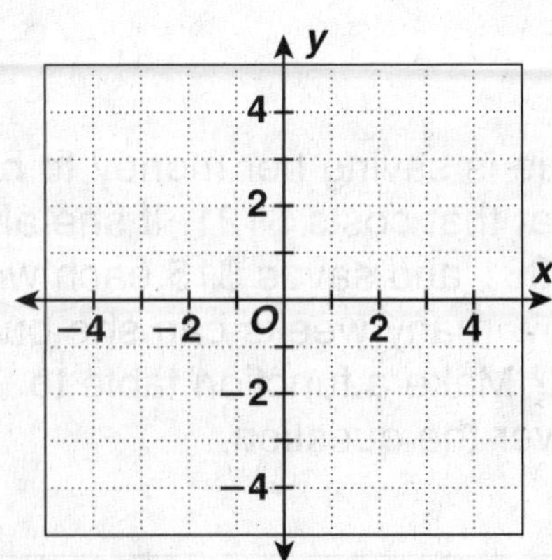

Holt Mathematics

 Challenge
Linear Functions

A linear equation is referred to as a "first-degree polynomial equation." The *standard form* of a first-degree equation is $ax + by + c = 0$. In standard form, a, b, and c are integers.

The *slope-intercept form* of a linear equation is $y = mx + b$. In slope-intercept form, m and b are both real numbers.

You can write a linear equation in either of these forms.

Example 1: Write $y = \frac{2}{3}x - 2$ in standard form.

$$y = \frac{2}{3}x - 2$$
$$y + 2 = \frac{2}{3}x \quad \longleftarrow \text{ Add 2 to both sides.}$$
$$3(y + 2) = 3\left(\frac{2}{3}x\right) \quad \longleftarrow \text{ Multiply both sides by 3.}$$
$$3y + 6 = 2x \quad \longleftarrow \text{ Simplify.}$$
$$-2x + 3y + 6 = 0 \quad \longleftarrow \text{ Subtract } 2x \text{ from both sides.}$$

Example 2: Write $3x + 5y + 1 = 0$ in slope-intercept form.
$$3x + 5y + 1 = 0$$
$$5y = -3x - 1 \quad \longleftarrow \text{ Subtract } 3x + 1 \text{ from both sides.}$$
$$\frac{5y}{5} = \frac{-3x - 1}{5} \quad \longleftarrow \text{ Divide both sides by 5.}$$
$$y = -\frac{3}{5}x - \frac{1}{5} \quad \longleftarrow \text{ Simplify.}$$

Write each equation in standard form.

1. $y = \frac{5}{6}x - 4$
2. $y = 2x - \frac{1}{2}$
3. $y = -\frac{1}{3}x + 8$

_________________   _________________   _________________

_________________   _________________   _________________

Write each equation in slope-intercept form.

4. $x - 4y + 1 = 0$
5. $2x - 4y + 8 = 0$
6. $6x - 3y - 2 = 0$

_________________   _________________   _________________

Holt Mathematics

Problem Solving
Graphing Linear Functions

Write the correct answer.

This graph shows the approximate population density (people per square mile) in the United States from 1950 to 2000.

1. Based on this graph, do you think the population of the United States is growing, shrinking, or staying about the same?

____

2. How do you know that this is the graph of a linear function?

____

____

3. On average, by about how many people did the density increase every ten years?

____

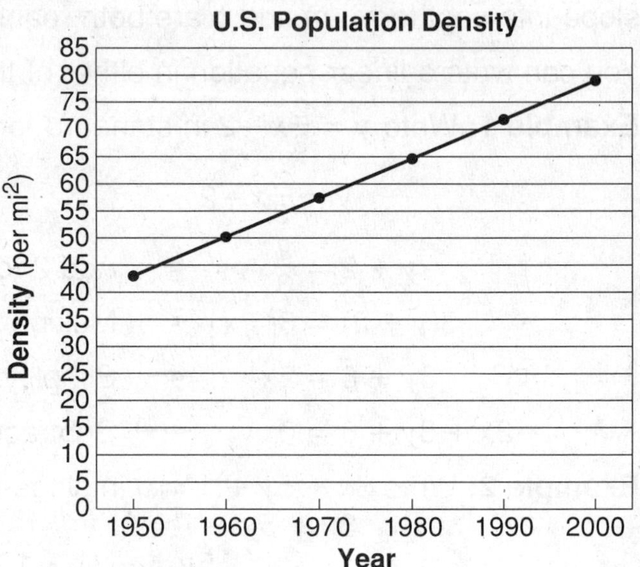

4. Estimate the population density in 2010.

____

5. Estimate the population density in 2050.

____

Choose the letter of the best answer.

6. Given the linear equation $y = 4x + 1$ and the input $x = 3$, what is the resulting ordered pair?

  A (15, 3)   **C** (12, 3)

  B (3, 13)   **D** (12, 11)

7. Given the linear equation $y = \frac{1}{3}x - 4$ and the input $x = 6$, what is the resulting ordered pair?

  F (6, −2)   **H** (−3, 6)

  G (6, 2)   **J** $\left(\frac{1}{3}, 6\right)$

8. A technician is adding chemicals to a 210-liter tank. She is adding the liquid at a rate of 2.5 liters per minute. How long will it take to fill the tank half full?

  A 42 min   **C** 168 min

  B 84 min   **D** 525 min

9. A basic set of 4 golf lessons costs $95.00. Additional lessons can be purchased at the discounted rate of $15.00 each. What will Veronica pay for a series of 10 lessons?

  F $150   **H** $185

  G $155   **J** $245

Holt Mathematics

Reading Strategies
Use a Graphic Organizer

This chart can help you understand the different ways to represent a
linear function.

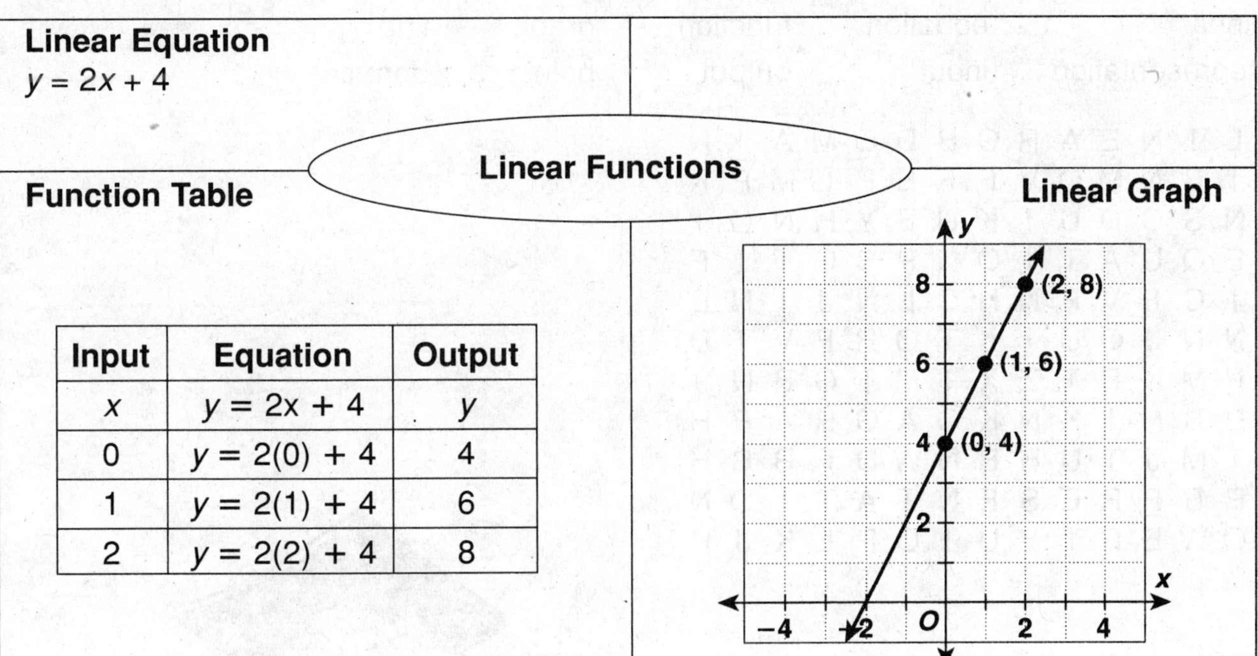

Use the graphic organizer to answer the following questions.

1. In addition to a linear equation, what are two other ways to
   represent a linear function?

   ____

2. Write the linear equation for the graph above.

   ____

3. For the input value 2, what is the output value?

   ____

4. For the output value 4, what is the input value?

   ____

5. How is the graph related to the function table?

   ____

   ____

Holt Mathematics

<table><tr><td>LESSON
4-6</td><td></td></tr></table>

Puzzles, Twisters & Teasers
Graduation Day!

Find and circle words from the list in the word search. Find a word that solves the riddle. Circle it and write it on the line.

linear	equation	function	graph	line
representation	input	output	point	domain

```
L I N E A R G B D O M A I N
I B N M O V F F E P L M P K
N S C D U I K U E Y H N O I
E Q U A T I O N P B G T I F
I C F V P N H C L N J I N B
N N J O U I L T O R F V T D
P M K P T Z X I M T G B N J
U B H I K N K O A G R A P H
T M J T U H B N W D F R B H
R E P R E S E N T A T I O N
Q W E R T Y U I O P L K J H
```

What does a fish get when it graduates from school?

A ___ ___ ___ ___ ___ ___ ___ ___ ___

Holt Mathematics

Practice A
The Coordinate Plane

Identify the quadrant of each point.

1. A   I
2. B   III
3. C   IV
4. D   II

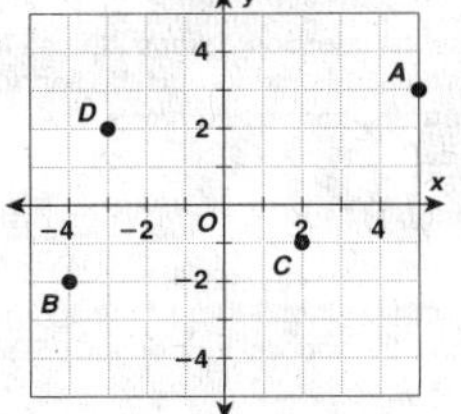

Plot each point.

5. (0, 3)
6. (−2, 4)
7. (5, 1)
8. (−4, −3)
9. (3, −2)
10. (−3, −1)

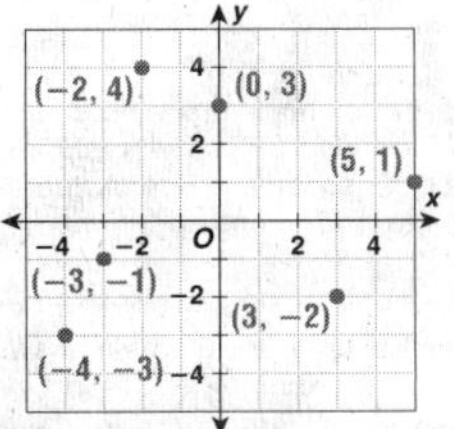

Give the coordinates of each point.

11. J   (−4, 1)
12. K   (0, −2)
13. L   (−3, −3)
14. M   (5, 2)
15. N   (3, −4)
16. P   (4, 0)
17. R   (2, −1)
18. S   (−1, 4)

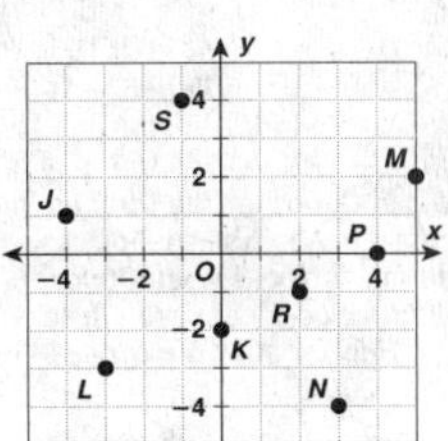

Holt Mathematics

Practice B
The Coordinate Plane

Identify the quadrant that contains each point.

1. A   III
2. B   IV
3. C   II
4. D   I

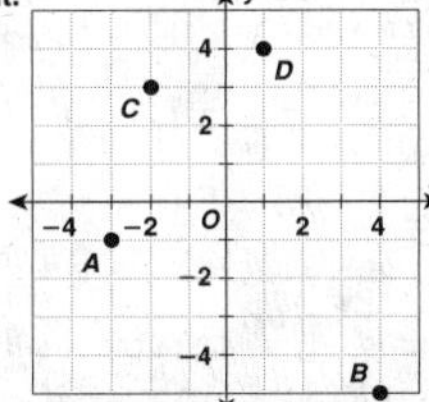

Plot each point on a coordinate plane.

5. (−4, 0)
6. (3, −3)
7. (1, 4)
8. (−5, −1)
9. (−2, 2)
10. (−1, −4)

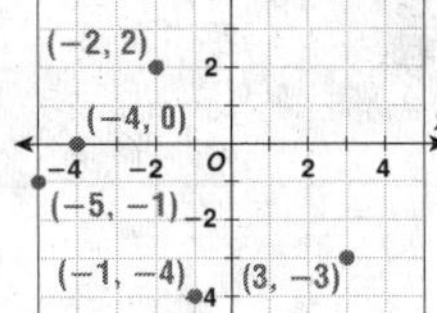

Give the coordinates of each point.

11. P   (−2, 5)
12. Q   (3, 0)
13. R   (4, −3)
14. S   (−5, −4)
15. T   (0, 2)
16. U   (−2, −2)
17. W   (3, 3)
18. X   (2, −1)

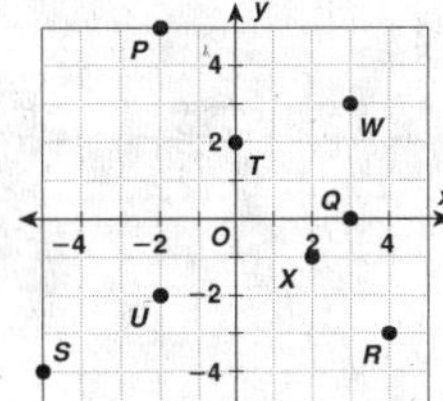

Holt Mathematics

Practice C
The Coordinate Plane

Identify the quadrant that contains each point.
Give the coordinates of each point.

1. A   II; (−2, 5)
2. B   IV; (3, −3)
3. C   I; (1, 4)
4. D   III; (−1, −2)

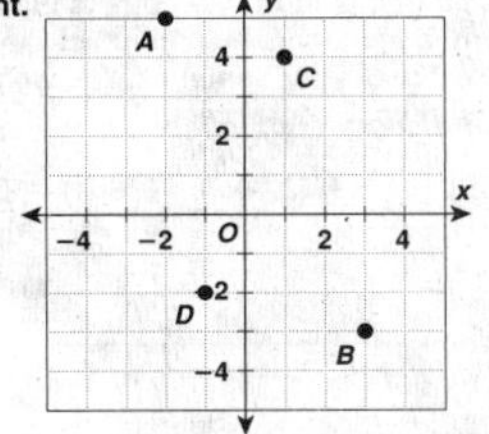

Plot each point on a coordinate plane.

5. (−2, −2)
6. (4, −5)
7. (0, −2)
8. (−4, 4)
9. (3, 0)
10. (−3, 1)

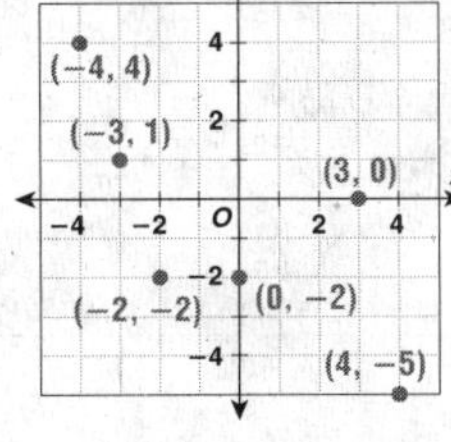

11. Graph the points (3, −2), (−3, −2) (−4, 3) and (2, 3). Connect each point in the order listed. Name the figure and the quadrants in which it is located.

parallelogram;
quadrants I, II, III, and IV

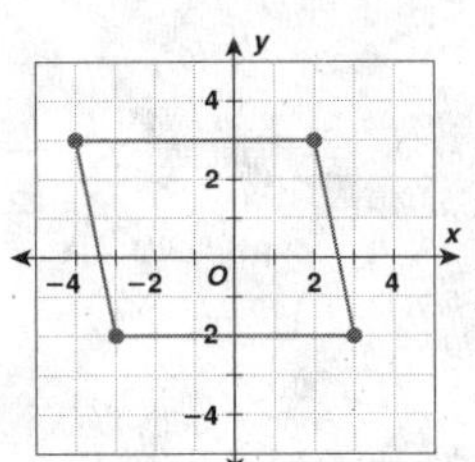

Holt Mathematics

Reteach
The Coordinate Plane

Numbers are graphed on a number line. **Ordered pairs** of numbers are graphed on a **coordinate plane.** A coordinate plane has two perpendicular number lines that divide it into **4 quadrants.** The following chart will help you identify the quadrants and the **coordinates** of points on a coordinate plane.

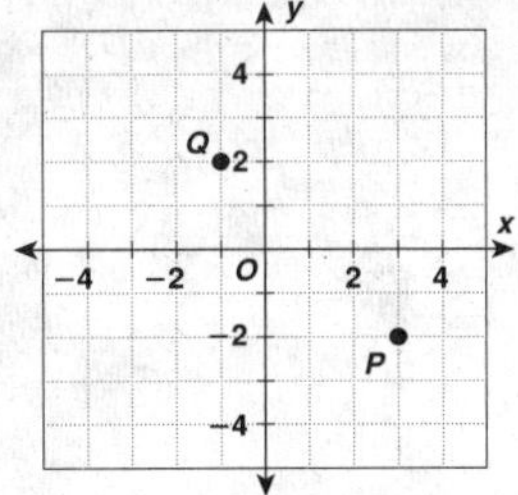

Quadrant II	Quadrant I
(−, +)	(+, +)
(←, ↑)	(→, ↑)

Quadrant III	Quadrant IV
(−, −)	(+, −)
(←, ↓)	(→, ↓)

To find the coordinates of P, start at (0, 0). Move 3 units →, then 2 units ↓. So the coordinates of P are (3, −2), and P is in quadrant IV.

To plot point Q with coordinates (−1, 2), start at (0, 0). Move 1 unit ←, then 2 units ↑. Q is in quadrant II.

Identify the quadrant and the coordinates of each point on the coordinate plane at the right.

1. A   II; (−4, 2)
2. B   I; (3, 4)
3. C   III; (−5, −2)
4. D   I; (1, 3)
5. E   IV; (1, −4)
6. F   III; (−2, −3)

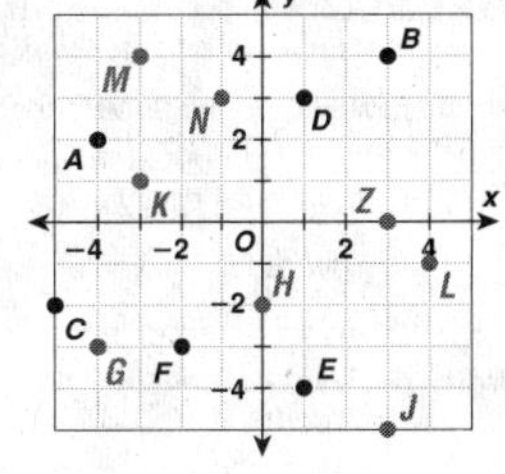

Plot each point on the coordinate plane above.

7. G (−4, −3)
8. H (0, −2)
9. J (3, −5)
10. K (−3, 1)
11. L (4, −1)
12. M (−3, 4)
13. N (−1, 3)
14. Z (3, 0)

Holt Mathematics

Holt Mathematics

 Challenge
Where in the World?

Graph each point on the grid below. Connect each point to the previous one as you graph it. Then connect the last point to the first point.

1. $(0, -10)$
2. $(-1, -9)$
3. $(-2.5, -7)$
4. $(-5, -7)$
5. $(-6, -5)$
6. $(-10, -5)$
7. $(-13, -3)$
8. $(-15, -1)$
9. $(-16, 2)$
10. $(-15, 8)$
11. $(-15, 10)$
12. $(-3, 9)$
13. $(4, 8)$
14. $(4, 7)$
15. $(6, 8)$
16. $(6, 4)$
17. $(8, 6)$
18. $(9, 6)$
19. $(9, 3)$
20. $(11, 5)$
21. $(16, 10)$
22. $(18, 8)$
23. $(16, 6)$
24. $(18, 4)$
25. $(14, 1)$
26. $(14, -1)$
27. $(11, -5)$
28. $(12.5, -8)$
29. $(13, -10)$
30. $(11, -9)$
31. $(9, -6)$
32. $(2.5, -7)$

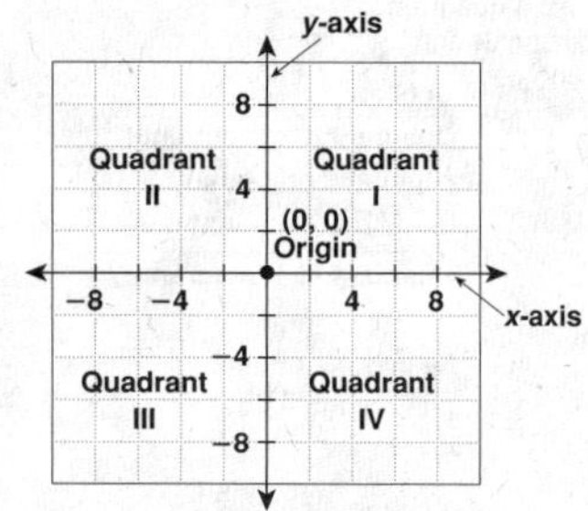

33. In which state is the point $(0, -8)$? **Texas**
34. Name a point in the state of Florida. **Possible Answer: (12, −9)**

7 **Holt Mathematics**

 Problem Solving
The Coordinate Plane

Write the correct answer.

1. Use the coordinate plane at right. In which quadrant(s) would the figure drawn by connecting points *J*, *K*, and *N* be?

   Quadrant IV

2. Use the coordinate plane at right. In which quadrant(s) would the figure drawn by connecting points *C*, *F*, and *M* be?

   Quadrants II and III

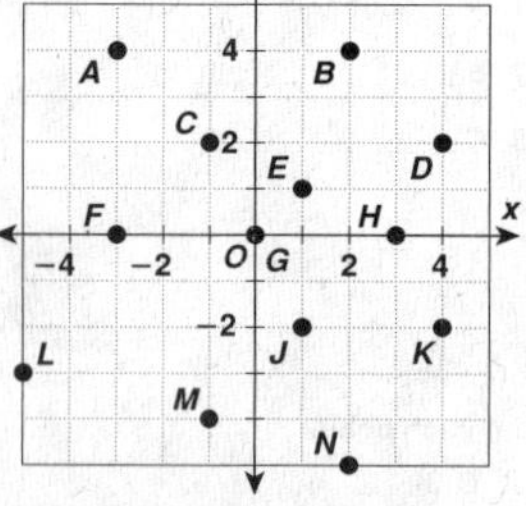

3. Maxine left home and walked 5 blocks north, 5 blocks west, 5 blocks south, and 5 blocks east. Where did she end up?

   back at home

4. Mr. Chin drove 2 miles north, then 3 miles east, then 2 miles south. How far is Mr. Chin from where he started?

   3 miles

Choose the letter for the best answer.

5. Which one of these points lies in Quadrant II of the coordinate plane above?
   - A $(5, 1)$
   - B $(5, -1)$
   - C $(-5, 1)$
   - D $(-5, -1)$

6. In which quadrant of the coordinate plane above is the figure formed by joining $(-4, -5)$, $(-2, -3)$ and $(-1, -1)$?
   - F Quadrant I
   - G Quadrant II
   - H Quadrant III
   - J Quadrant IV

7. Abe and Carlos left the library at the same time. Abe walked 4 blocks north and 5 blocks west. Carlos walked 4 blocks east and 4 blocks north. How far apart were they?
   - A 10 blocks
   - B 9 blocks
   - C 8 blocks
   - D 5 blocks

8. When a point lies on the *x*-axis, which of these must be true?
   - F The *x*-coordinate is 0.
   - G The *y*-coordinate is 0.
   - H The *x*-coordinate is greater than the *y*-coordinate.
   - J The *y*-coordinate is greater than the *x*-coordinate.

8 **Holt Mathematics**

 Reading Strategies
Using Relevant Information

Horizontal number lines and vertical number lines form a grid called the **coordinate plane**.

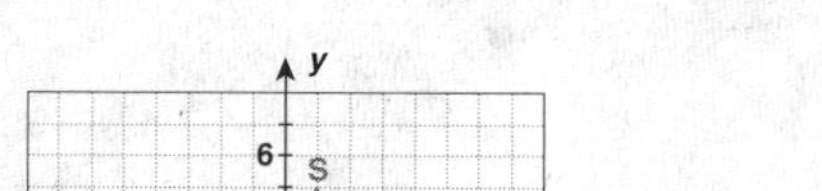
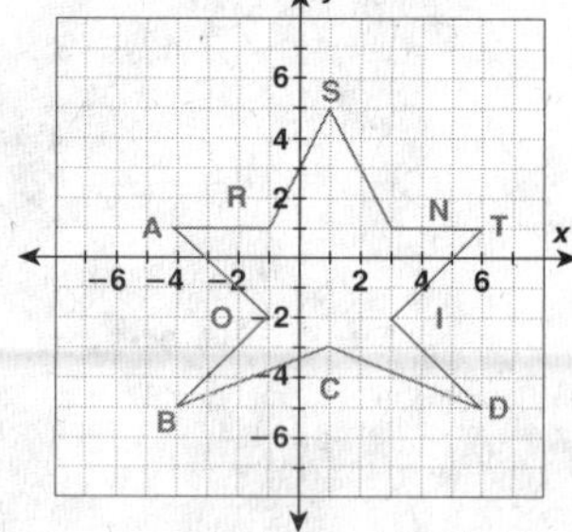

1. What is the name of the horizontal number line? **x-axis**
2. What is the name of the vertical number line? **y-axis**
3. The number lines meet at point $(0, 0)$. What is that point called? **the origin**
4. What are the four parts that divide the coordinate plane called? **quadrants**

To locate a point on the coordinate plane, you always start at the origin. You first move either to the right or left along the **x-axis**.

Write "positive" or "negative" to show which direction you are moving from zero.

5. If you move to the right, you are moving in a **positive** direction.
6. If you move to the left, you are moving in a **negative** direction.

From your position on the *x*-axis, you move up or down along the **y-axis**.

Write "positive" or "negative" to show which direction you are moving from zero.

7. If you move up, you are moving in a **positive** direction.
8. If you move down, you are moving in a **negative** direction.

9 **Holt Mathematics**

 Puzzles, Twisters & Teasers
Plane Thinking!

Plot the points in the coordinate plane below. Fill in the blanks to form a sentence, then connect the points to make a picture that matches your sentence. You won't need to use all the letters in your sentence.

S $(1, 5)$ N $(3, 1)$ T $(6, 1)$ R $(-1, 1)$ A $(-4, 1)$
O $(-1, -2)$ I $(3, -2)$ B $(-4, -5)$ D $(6, -5)$ C $(1, -3)$

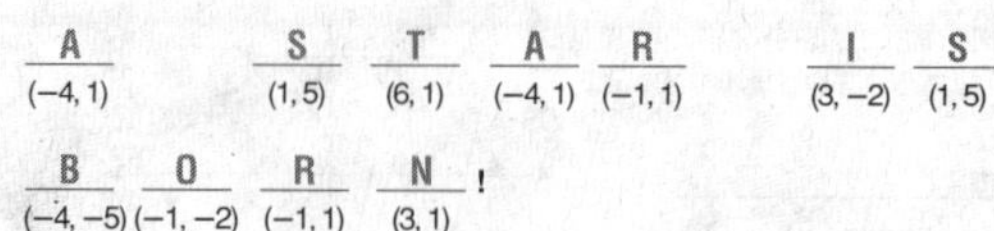

$\underset{(-4,1)}{A}$ $\underset{(1,5)}{S}$ $\underset{(6,1)}{T}$ $\underset{(-4,1)}{A}$ $\underset{(-1,1)}{R}$ $\underset{(3,-2)}{I}$ $\underset{(1,5)}{S}$

$\underset{(-4,-5)}{B}$ $\underset{(-1,-2)}{O}$ $\underset{(-1,1)}{R}$ $\underset{(3,1)}{N}$!

10 **Holt Mathematics**

---

52 **Holt Mathematics**

Practice A
Tables and Graphs

Write each ordered pair from the table.

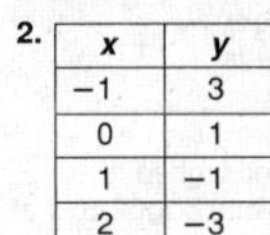

1.

x	y		(x, y)
2	4	→	(2, 4)
3	5	→	(3, 5)
4	6	→	(4, 6)
5	7	→	(5, 7)

Graph the ordered pairs from the table.

2.

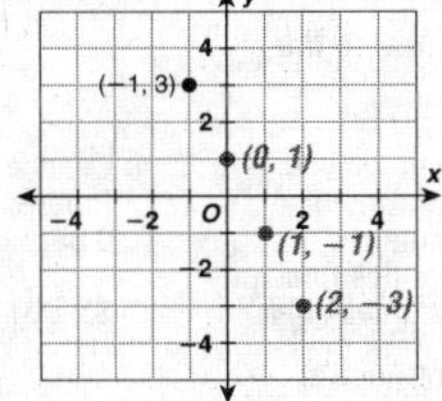

x	y
−1	3
0	1
1	−1
2	−3

3. The table shows the cost for different weights of bananas. Graph the data.

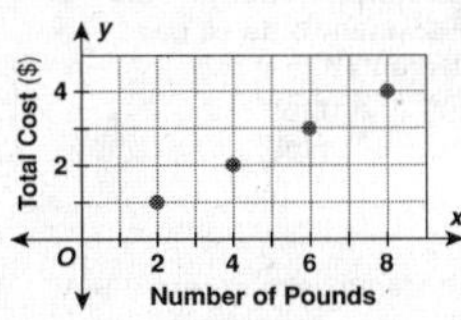

Pounds of Bananas	Total Cost
2	$1
4	$2
6	$3
8	$4

What appears to be the relationship between the number of pounds and total cost?

<u>The total cost increases as the number of pounds increases.</u>

Holt Mathematics

Practice B
Tables and Graphs

Write each ordered pair from the table.

1.

x	y		(x, y)
−12	−8	→	(−12, −8)
−9	−4	→	(−9, −4)
−6	0	→	(−6, 0)
−3	4	→	(−3, 4)

Graph the ordered pairs from the table.

2.

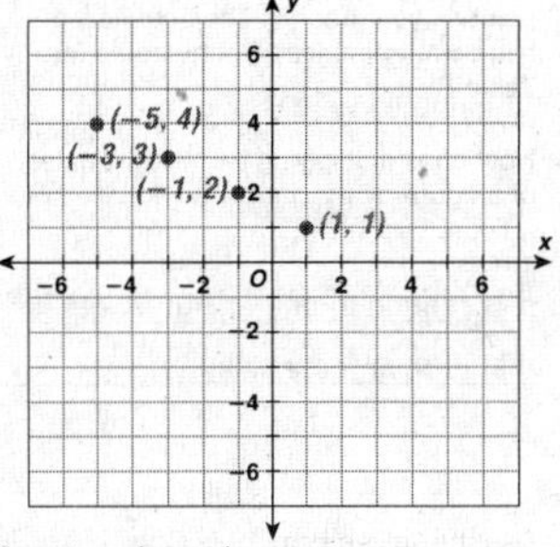

x	y
−5	4
−3	3
−1	2
1	1

3. The table shows the cost of buying different numbers of souvenir pens. Graph the data.

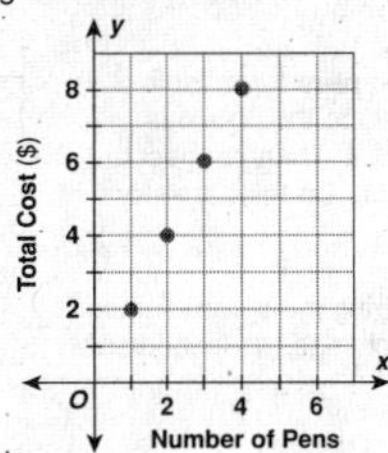

Number of Pens	Total Cost
1	$2
2	$4
3	$6
4	$8

What appears to be the relationship between the number of pens and total cost?

<u>The total cost increases as the number of pens increases.</u>

Holt Mathematics

Practice C
Tables and Graphs

Write each ordered pair from the table.

1.

x	y		(x, y)
−7	12	→	(−7, −12)
−4	10	→	(−4, 10)
−1	8	→	(−1, 8)
2	6	→	(2, 6)

Graph the ordered pairs from the table.

2.

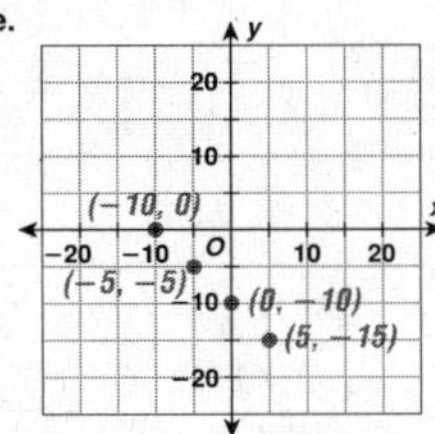

x	y
−10	0
−5	−5
0	−10
5	−15

3. A 40-gallon water tank that was full developed a leak.

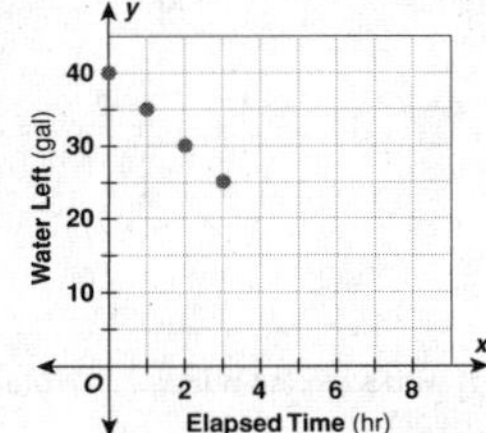

Elapsed Time (hr)	Water Remaining (gal)
0	40
1	35
2	30
3	25

a. Make a graph of the data.

b. Explain how you can use the graph to find the amount of water remaining after 6 hours.

<u>Possible answer: Draw a line through the points. Extend the line until it reaches the point at which the x-value (hours) is equal to 6.</u>

Holt Mathematics

Reteach
Tables and Graphs

You can write an ordered pair for each pair of values in a table of values.
Write the x value as the first number in the ordered pair.
Write the y value as the second number in the ordered pair.

Complete the missing values in the ordered pairs.

	x	y		Ordered Pairs (x, y)
	−2	−4	→	(−2, −4)
1.	0	−3	→	(0, −3)
2.	2	−2	→	(2, −2)
3.	4	−1	→	(4, −1)

You can use the ordered pairs to graph the pairs of values in the table.

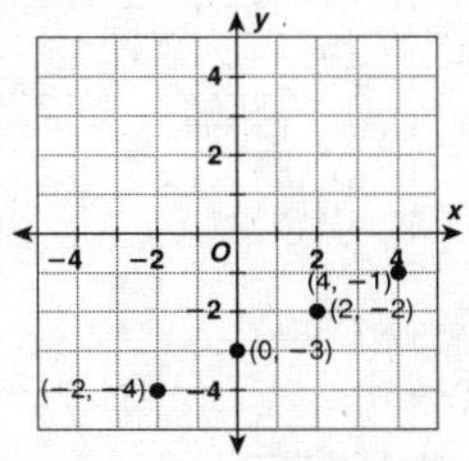

Write the ordered pairs from the table. Then graph the ordered pairs.

4.

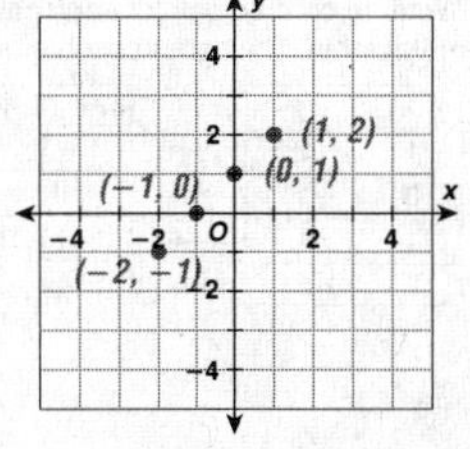

x	y		(x, y)
−2	−1	→	(−2, −1)
−1	0	→	(−1, 0)
0	1	→	(0, 1)
1	2	→	(1, 2)

Holt Mathematics

Holt Mathematics

Challenge
LESSON 4-2
Patterns on a Coordinate Plane

You can graph points and shapes on a coordinate plane. If you multiply the coordinates of these points by negative numbers, you will find patterns in the results.

1. Use the values from Table A to complete Table B.

2. Graph the ordered pairs from Table A. Connect each point to the previous point as you graph it. Then connect the last point to the first. Do the same for Table B.

Table A

x	y
1	3
2	1
3	2
4	4

Table B

−x	y
−1	3
−2	1
−3	2
−4	4

3. How does multiplying the x-coordinates of a figure by −1 change the location of a figure?

It is reflected, or a mirror image is made, over the y-axis.

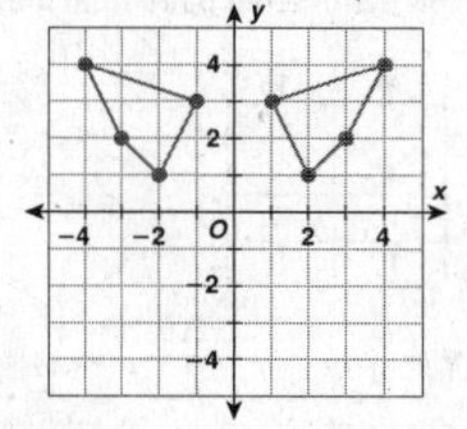

4. Use the values from Table C to complete Table D.

5. Graph the ordered pairs from Table C. Connect each point to the previous point as you graph it. Then connect the last point to the first. Do the same for Table D.

Table C

x	y
1	2
2	4
3	3
4	1

Table D

x	−y
1	−2
2	−4
3	−3
4	−1

6. How does multiplying the y-coordinates of a figure by −1 change the location of a figure?

It is reflected, or a mirror image is made, over the x-axis.

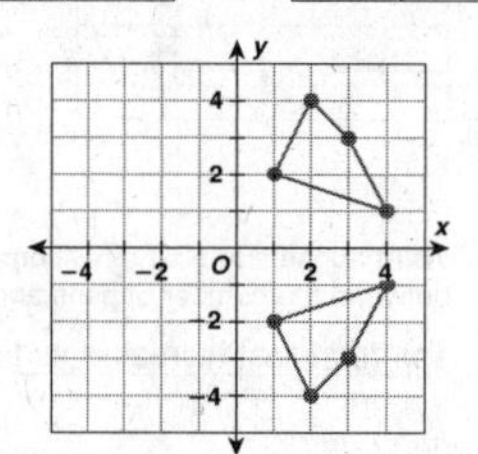

15   **Holt Mathematics**

Problem Solving
LESSON 4-2
Tables and Graphs

Write the correct answer.

1. The table shows the total cost of buying different numbers of bags of peanuts. Graph the data.

Number of Bags	1	2	3	4
Total Cost	$0.50	$1.00	$1.50	$2.00

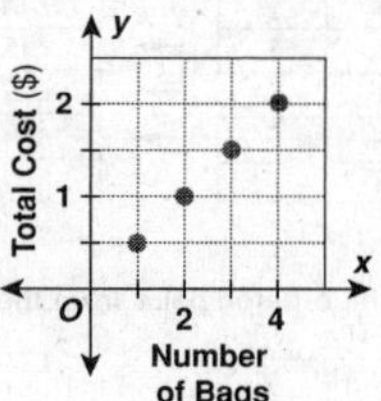

2. What appears to be the relationship between the number of bags of peanuts and total cost shown in Problem 1?

The total cost increases as the number of bags increases.

3. How could you use the graph you made in Problem 1 to predict the total cost of 6 bags of peanuts?
Possible answer: Draw a line through the points and extend it to the point at which the number of bags is 6.

4. Use the plan you described in Problem 3 to predict the total cost of 6 bags of peanuts.

$3.00

Choose the letter for the best answer.

5. Terry plotted the ordered pairs from the table. How many ordered pairs were in Quadrant I?

x	y
−4	5
−1	3
2	1
5	−1

A 0   C 2
(B) 1   D 3

6. How many ordered pairs were in Quadrant II?

F 0   (H) 2
G 1   J 3

7. Which ordered pair is in Quadrant IV?

A (−4, 5)   C (2, 1)
B (−1, 3)   (D) (5, −1)

16   **Holt Mathematics**

Reading Strategies
LESSON 4-2
Multiple Representations

You can represent pairs of x-values and y-values in three different ways.

Way 1: Make a **table of values.**

x	y
−5	−1
−2	1
1	3
4	5

Way 2: Write **ordered pairs.** Use the form (x, y).

(x, y)
(−5, −1)
(−2, 1)
(1, 3)
(4, 5)

Way 3: Graph ordered pairs on a coordinate plane. Start at the origin. Move left or right for the x-value. Move up or down for the y-value

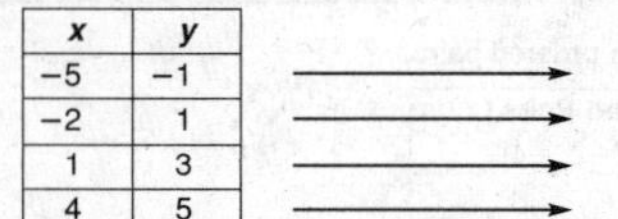

1. In the ordered pair (3, 5), which number is the x-value? **3**

2. In the ordered pair (4, −8), which value is the y-value? **−8**

3. How can you use a graph to represent the ordered pair (−2, 5)?

Start at the origin; move 2 units left; then move 5 units up.

4. Write an ordered pair for each pair of x- and y-values in the table. Then graph each ordered pair.

x	y		(x, y)
−1	−3	→	(−1, −3)
0	−1	→	(0, −1)
1	1	→	(1, 1)
2	3	→	(2, 3)

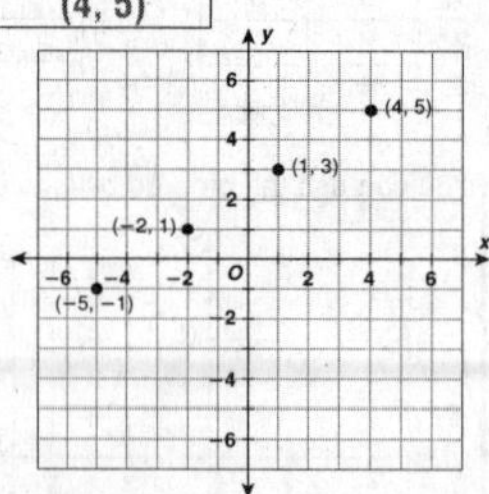

17   **Holt Mathematics**

Puzzles, Twisters & Teasers
LESSON 4-2
I Spy!

Find the point for each ordered pair on the coordinate plane. Write the letter that names each point.

	x	y	
1.	−5	4	S
2.	−3	1	P
3.	−1	−2	Y
4.	1	−5	F

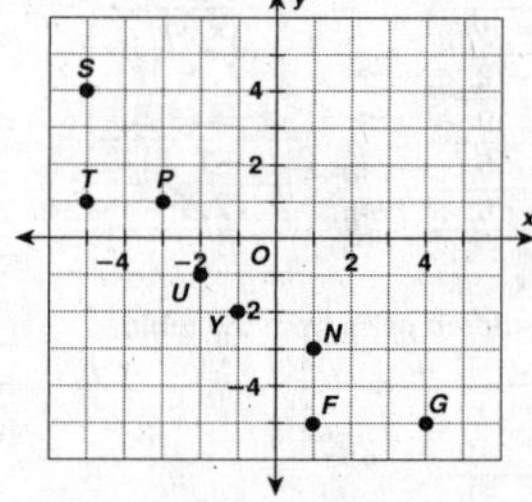

	x	y	
5.	−6	−4	O
6.	−4	−2	C
7.	−2	0	A
8.	0	2	L
9.	2	4	S

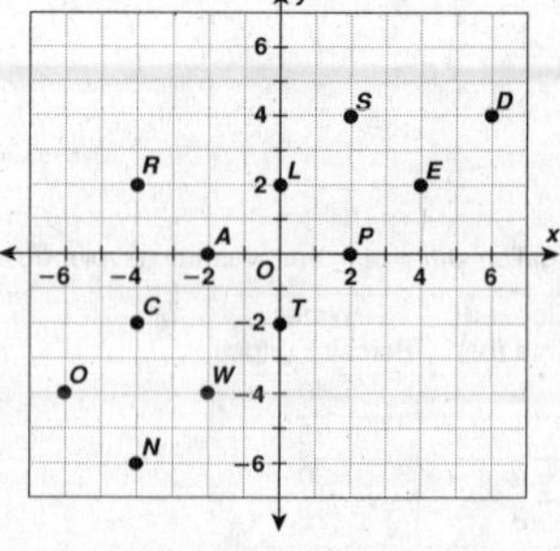

Start with exercise number 1. Write the letters in order to solve the riddle.

What kind of glasses do secret agents wear?

S P Y . F O C A L S

18   **Holt Mathematics**

Holt Mathematics

Practice A
Interpreting Graphs

Match each graph with the description of the relationship it shows.

1. heating water to the boiling point

   Graph __C__

2. cooling water to the freezing point

   Graph __A__

3. pulling the plug in the bathtub

   Graph __B__

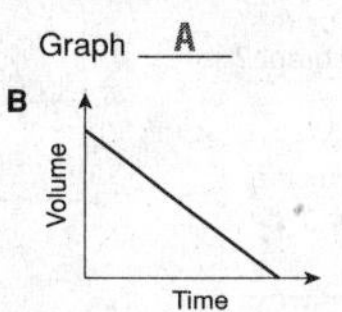 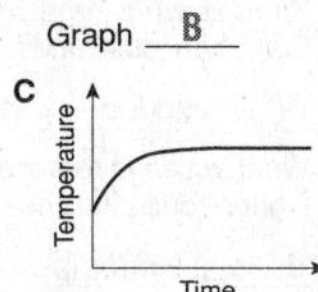

A 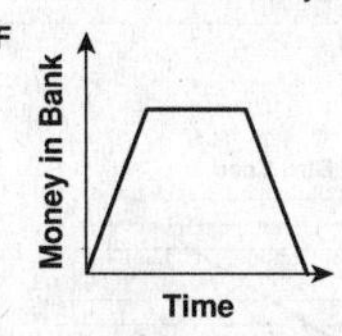   **B** 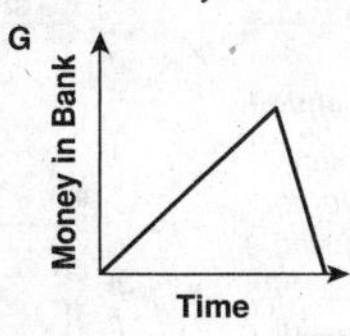   **C** 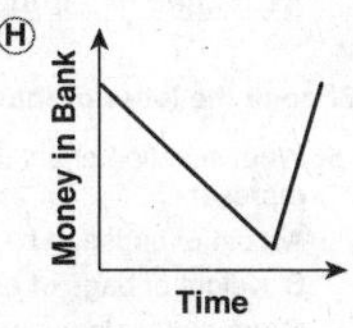

4. Ben withdrew money from his bank account each week for 6 weeks. Then he deposited money into the account. Which graph best shows the story? Circle the letter of your answer.

F   **G**   **(H)**

5. Talia hiked for 4 miles, and stopped to eat lunch. Then she hiked back to where she started. Complete the graph. Show the total distance Talia hiked compared to the time that she took.

6. Use your graph to find the total number of miles Talia hiked.

   __8 miles__

19
Holt Mathematics

Practice B
Interpreting Graphs

1. The gas tank in Karen's car was full. Karen drove the car until only $\frac{1}{4}$ of the tank was full. Karen filled up the tank again and drove the car until $\frac{1}{4}$ of the tank was full. Which graph best shows the story? Circle the letter of your answer.

A 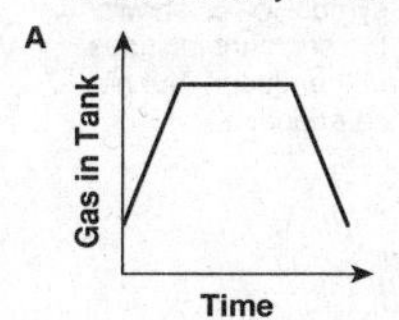   **(B)** 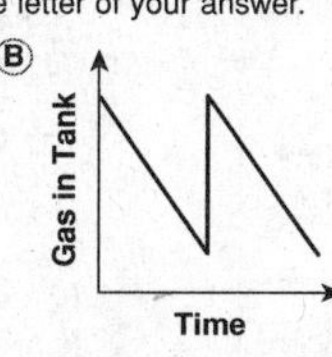   **C** 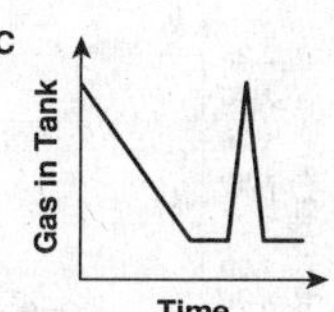

2. An elevator started at the ground floor. It went up to the sixth floor and stopped, then went to the fourth floor and stopped, and finally returned to the ground floor. Which graph best shows the story? Circle the letter of your answer.

(F) 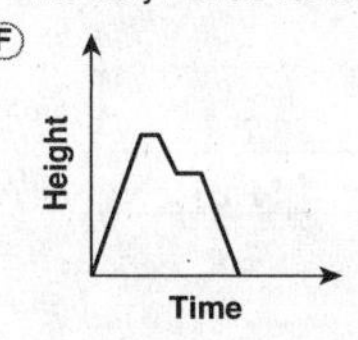   **G** 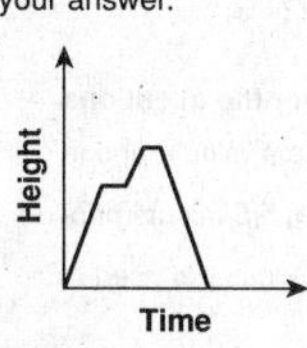   **H** 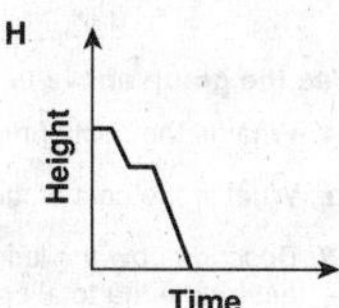

3. Maxine biked 6 miles from her house to the park. She played some softball. Then she biked 4 miles farther to the movie theater. After watching a movie, Maxine returned home. Complete the graph so that it shows the distance Maxine is from home compared to the time.

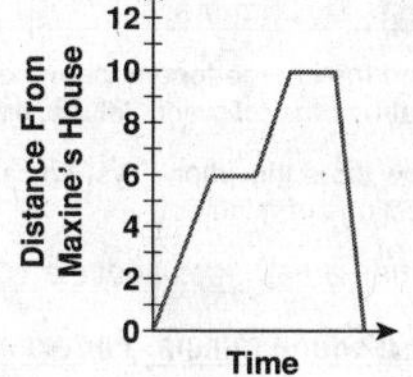

4. Use your graph to find the total number of miles Maxine biked.

   __20 miles__

20
Holt Mathematics

Practice C
Interpreting Graphs

1. An elevator was on the tenth floor of an office building. It came down and stopped at the fourth floor. It continued down, stopping at each of the remaining floors until it reached the first floor. Which graph best shows the story? Circle the letter of your answer.

A 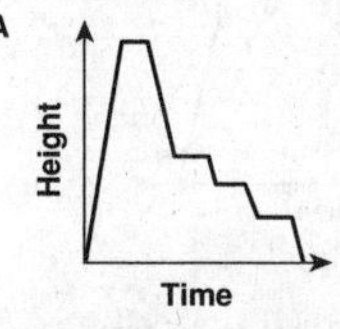   **B**   **(C)**

2. Yolanda took 16 weeks to save money for a trip. For the first 8 weeks, she saved $50 each week. For the next 4 weeks Yolanda did not save any money. Then Yolanda saved $50 a week for 4 weeks. When Yolanda took the trip, she spent $500 of her savings. Complete the graph so that it shows Yolanda's savings compared to time.

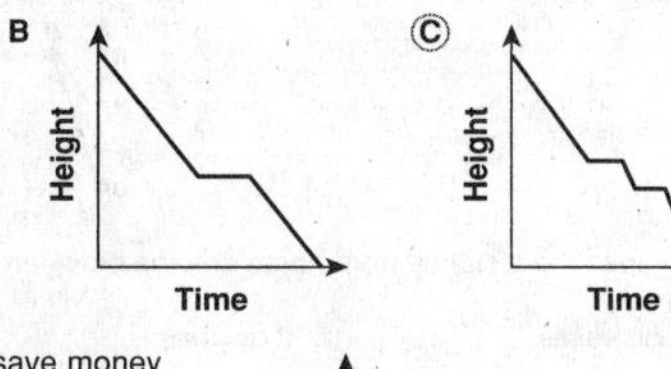

3. Use your graph to find the amount of money Yolanda had left after the trip.

   __$100__

Write a story that describes the graph below.

4. The location of a skier on a ski slope __Possible answer: A skier skis down the slope, waits for the chair lift, goes back up, skis down, waits for the chair lift, and goes back up again.__

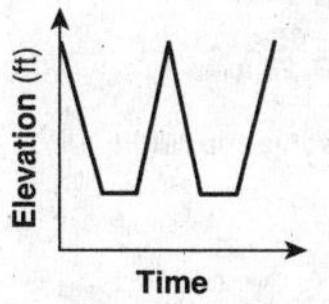

21
Holt Mathematics

Reteach
Interpreting Graphs

Graphs are often used to model situations. This graph shows Shavawn's daily jogging routine. She jogs uphill at a steady speed. When she starts to run downhill, her speed increases.

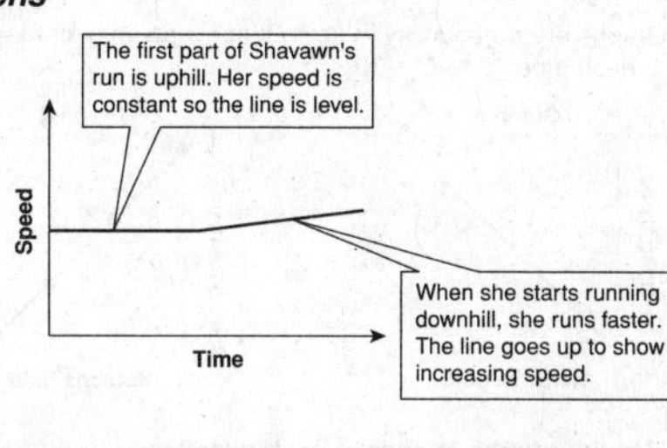

Complete the graph for each situation.

1. You pour warm tea into a glass, then add ice cubes.

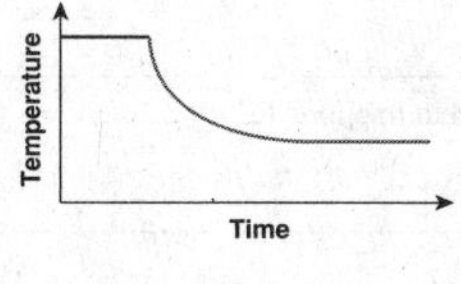

2. You are watching TV and lower the volume during a commercial.

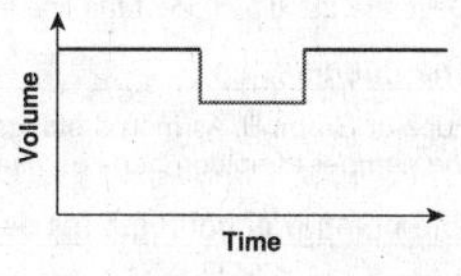

3. You bike from your home to the park. You play softball. Then you bike home.

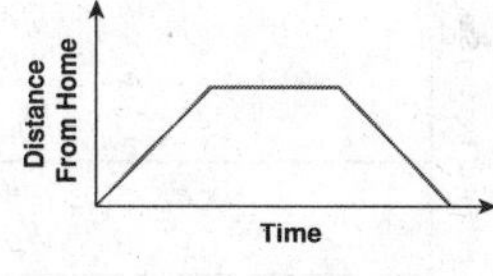

4. You drive at a steady speed. You hit a traffic jam, slow down, and travel at a slower speed. Then the traffic jam ends, and you travel at a faster speed.

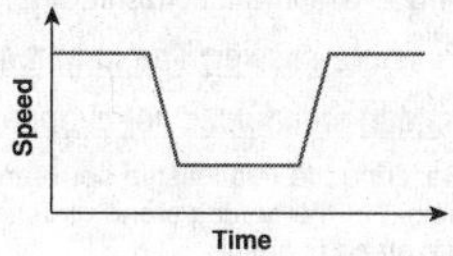

22
Holt Mathematics

Challenge
Step Functions

A step graph is a graph that looks like steps. You can use step graphs to represent certain kinds of relationships. The step graph below shows the total cost of calls of different lengths on a particular pay phone system.

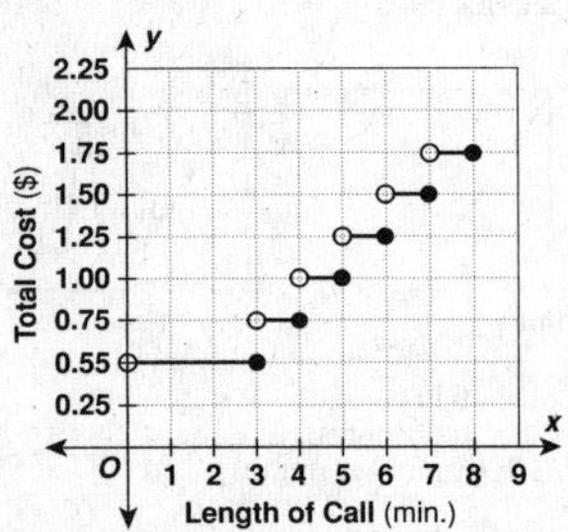

The symbol o—● shows that the segment includes the right endpoint, but not the left endpoint.

Use the graph above to answer the questions.

1. What is the cost of making a 4-minute phone call? **$0.75**

2. What is the cost of making a 5.5-minute phone call? **$1.25**

3. Describe how the length of the call is used to determine the total cost.

   A call of up to 3 minutes costs $0.50. Each additional minute costs $0.25 more.

4. Find the charge for any phone call with a length (l) that has the following values: 6 minutes < l ≤ 7 minutes. **$1.50**

5. How does this phone system calculate charges for fractions of minutes?

   When a call goes above a whole number of minutes, it charges for the next whole minute. For example, all calls that are greater than 6 minutes and less than or equal to 7 minutes have the same total charge.

23   **Holt Mathematics**

Problem Solving
Interpreting Graphs

Write the correct answer.

Eduardo exercises by doing laps around a track. He sprints the straight-aways and walks the curves.

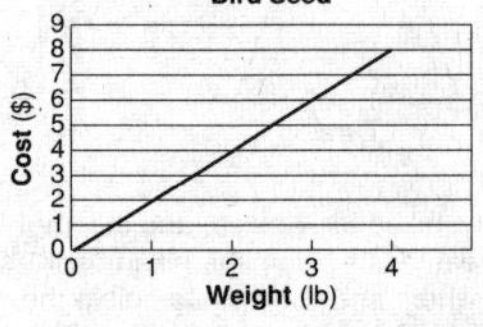

1. What action is represented by the lower horizontal lines in the graph?

   He is walking.

2. What action is represented by the higher horizontal lines in the graph?

   He is sprinting at full speed.

3. What action is represented by the dashed lines in the graph?

   He begins to sprint.

4. What action is represented by the dotted lines in the graph?

   He slows down.

Choose the letter of the best answer.

5. What situation could this graph represent?
   - **A** cost of birdseed by the pound
   - B weight of bags of birdseed
   - C ages of birds eating birdseed
   - D a buy-one-get-one-free sale

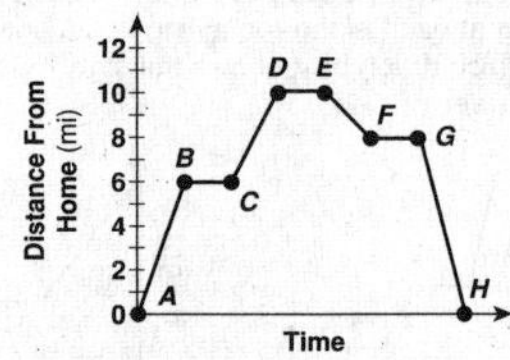

6. What would you expect to pay if you needed $3\frac{1}{2}$ pounds of birdseed?
   - F $3.50
   - G $5.00
   - **H** $7.00
   - J $7.50

7. What is the price for each pound of birdseed?
   - A $0.25
   - B $0.50
   - C $1.00
   - **D** $2.00

24   **Holt Mathematics**

Reading Strategies
Read a Graph

Graphs are a good way to show that events may or may not relate to each other.

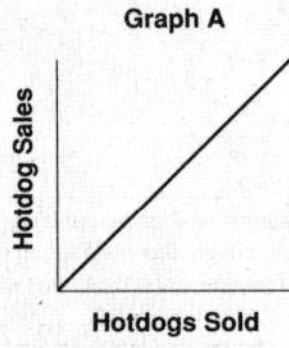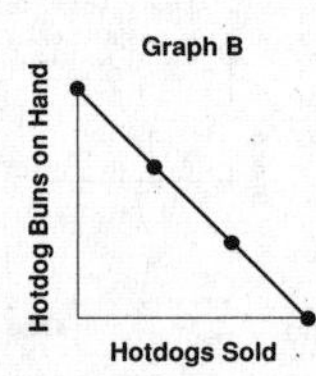

Use the graphs to answer each question.

1. Look at Graph A. As the number of hotdogs sold increases, what happens to hotdog sales?

   Hotdog sales increase.

2. Describe the direction of the line in Graph A.

   The line goes up.

3. Look at Graph B. As more hotdogs are sold, what happens to the number of hotdog buns on hand?

   The number of hotdog buns decreases.

This graph compares the height of a hotdog stand with the number of hours the hotdog stand is open. Use the graph to answer each question.

4. Why is Graph C a horizontal line?

   Possible answer: The height of the hotdog stand does not change.

5. Describe the relationship between the height of the hotdog stand and the hours the stand is open.

   There is none.

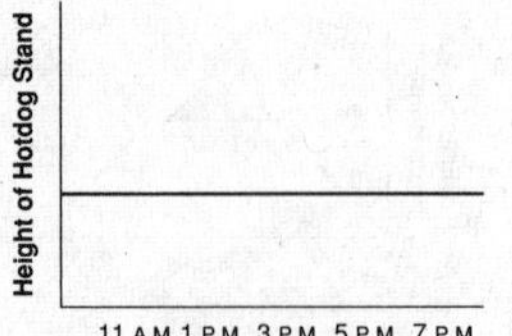

25   **Holt Mathematics**

Puzzles, Twisters, & Teasers
Graphically Speaking!

Use the graph to answer each question. Circle the letter next to your answer.

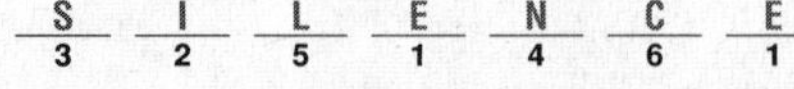

1. How does the distance from home change between points C and D?
   - **E** It increases.
   - S It decreases.
   - R It stays the same.

2. Which pair of points are the endpoints of a segment that represents an interval of time in which Jenny's distance from home does not change?
   - F Points A and B
   - **I** Points D and E
   - N Points E and F

3. At point B, Jenny is at a flower store. How many miles from Jenny's home is the flower store?
   - M 10 miles
   - R 8 miles
   - **S** 6 miles

4. When Jenny is furthest from home, how far from home is she?
   - A 8 miles
   - **N** 10 miles
   - F 12 miles

5. Which of the following could describe what Jenny is doing between points G and H?
   - T Jenny is traveling 8 miles to reach a store.
   - O Jenny is stopped at a store that is 8 miles from home.
   - **L** Jenny is traveling 8 miles to return home.

6. Suppose Jenny's home and the points at which she stops along the way are all along the same road. What is the total distance that Jenny travels?
   - W 18 miles
   - **C** 20 miles
   - G 28 miles

Write the circled letters above the problem numbers to solve the riddle.

What is broken once you've spoken?

S	I	L	E	N	C	E
3	2	5	1	4	6	1

26   **Holt Mathematics**

Holt Mathematics

Find the output for each input.

1. $y = x + 5$

Input	Rule	Output
x	x + 5	y
0	0 + 5	5
1	1 + 5	6
2	2 + 5	7

2. $y = 3x$

Input	Rule	Output
x	3x	y
0	3(0)	0
1	3(1)	3
2	3(2)	6

3. $y = 3x - 4$

Input	Rule	Output
x	3x − 4	y
0	3(0) − 4	−4
1	3(1) − 4	−1
2	3(2) − 4	2

4. $y = x^2$

Input	Rule	Output
x	x^2	y
0	$(0)^2$	0
1	$(1)^2$	1
2	$(2)^2$	4

Make a function table. Graph the resulting ordered pairs.

5. $y = 2x - 4$

Input	Rule	Output	Ordered Pair
x	2x − 4	y	(x, y)
0	2(0) − 4	−4	(0, −4)
1	2(1) − 4	−2	(1, −2)
2	2(2) − 4	0	(2, 0)

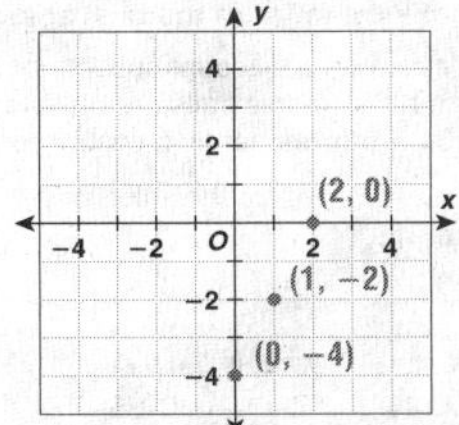

6. $y = 4x^2$

Input	Rule	Output	Ordered Pair
x	$4x^2$	y	(x, y)
−1	$4(-1)^2$	4	(−1, 4)
0	$4(0)^2$	0	(0, 0)
1	$4(1)^2$	4	(1, 4)

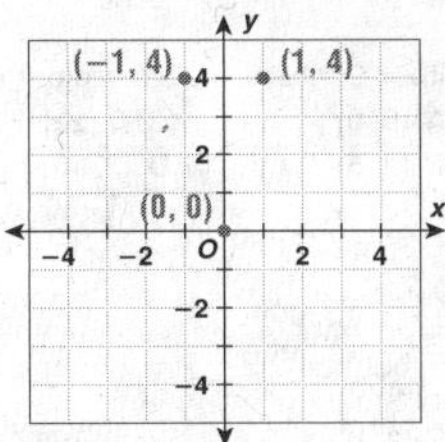

Holt Mathematics

Find the output for each input.

1. $y = 5x - 1$

Input	Rule	Output
x	5x − 1	y
−2	5(−2) − 1	−11
0	5(0) − 1	−1
3	5(3) − 1	14
6	5(6) − 1	29

2. $y = -2x^2$

Input	Rule	Output
x	$-2x^2$	y
−2	$-2(-2)^2$	−8
2	$-2(2)^2$	−8
3	$-2(3)^2$	−18
4	$-2(4)^2$	−32

Make a function table, and graph the resulting ordered pairs.

3. $y = x \div 4$

Input	Rule	Output	Ordered Pair
x	x ÷ 4	y	(x, y)
−4	−4 ÷ 4	−1	(−4, −1)
0	0 ÷ 4	0	(0, 0)
2	2 ÷ 4	$\frac{1}{2}$	$(2, \frac{1}{2})$
4	4 ÷ 4	1	(4, 1)

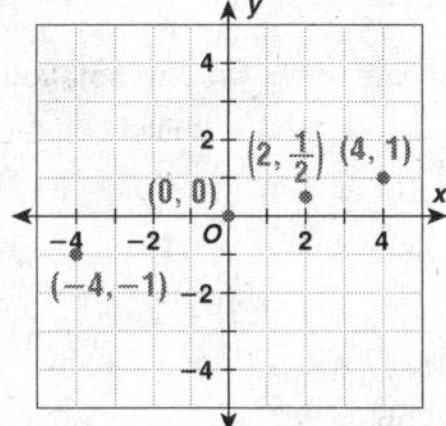

4. $y = x^2 - 5$

Input	Rule	Output	Ordered Pair
x	$x^2 - 5$	y	(x, y)
−2	$(-2)^2 - 5$	−1	(−2, −1)
−1	$(-1)^2 - 5$	−4	(−1, −4)
0	$(0)^2 - 5$	−5	(0, −5)
1	$(1)^2 - 5$	−4	(1, −4)

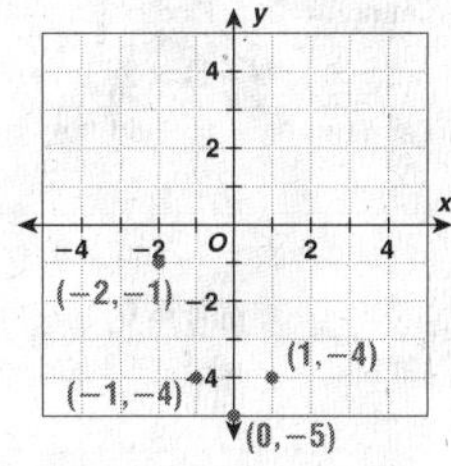

Holt Mathematics

Make a function table, and graph the resulting ordered pairs.

1. $y = \frac{x}{3}$

Input	Rule	Output	Ordered Pair
x	$\frac{x}{3}$	y	(x, y)
−3	$\frac{-3}{3}$	−1	(−3, −1)
0	$\frac{0}{3}$	0	(0, 0)
3	$\frac{3}{3}$	1	(3, 1)

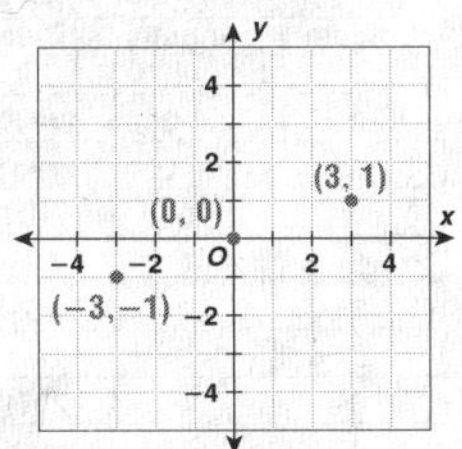

2. $y = 2x^2 - 4$

Input	Rule	Output	Ordered Pair
x	$2x^2 - 4$	y	(x, y)
−2	$2(-2)^2 - 4$	4	(−2, 4)
0	$2(0)^2 - 4$	−4	(0, −4)
1	$2(1)^2 - 4$	−2	(1, −2)
2	$2(2)^2 - 4$	4	(2, 4)

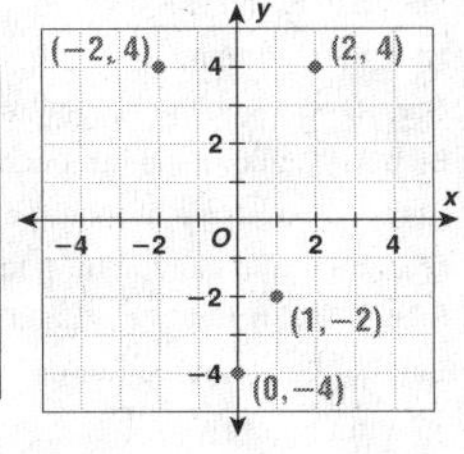

3. For every dollar you spend at the Rusty Nail hardware chain, you receive 20 Gold Hammer points that can be redeemed for free merchandise. The equation $y = 20x$ gives the number of points y received for spending x dollars. Make an input/output table using the values $x = 10, 25, 50, 80$.

Input	Rule	Output
x	20x	y
10	20(10)	200
25	20(25)	500
50	20(50)	1,000
80	20(80)	1,600

Holt Mathematics

A **function** is a relationship in which the value of one quantity depends on the value of another quantity. A function can be represented by a rule or an equation. A *function table* can help you find the ordered pair values for a function.

Complete the missing values in the table.

	Input	Rule	Output	Ordered Pair
	x	3x + 2	y	(x, y)
1.	−2	3(−2) + 2 = −6 + 2	−4	(−2, −4)
2.	−1	3(−1) + 2 = −3 + 2	−1	(−1, −1)
3.	0	3(0) + 2 = 0 + 2	2	(0, 2)
4.	1	3(1) + 2 = 3 + 2	5	(1, 5)

You can graph the resulting ordered pairs from the input/output table.

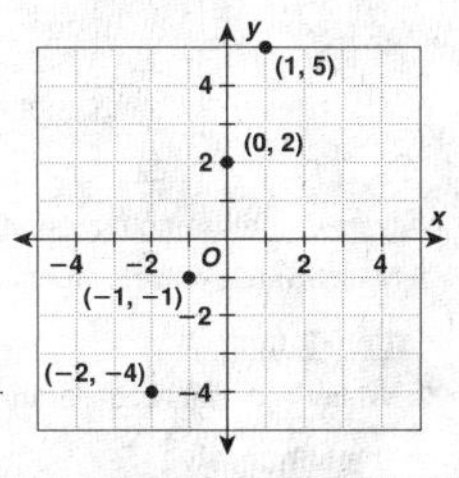

Find the output for each input value, and graph the resulting ordered pairs.

5. $y = 2x - 1$

Input	Rule	Output	Ordered Pair
x	2x − 1	y	(x, y)
−2	2(−2) − 1	−5	(−2, −5)
−1	2(−1) − 1	−3	(−1, −3)
0	2(0) − 1	−1	(0, −1)
1	2(1) − 1	1	(1, 1)
2	2(2) − 1	3	(2, 3)

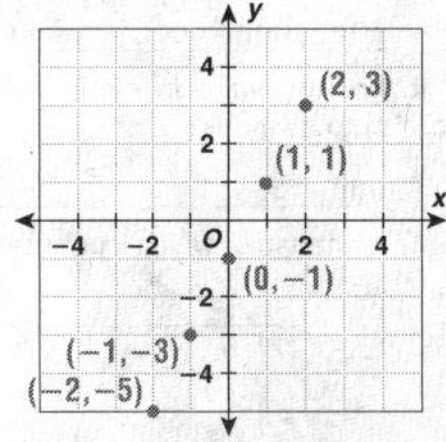

Holt Mathematics

Holt Mathematics

Challenge
Function Notation

You can use the equation of a function to find values for y that satisfy the equation for any value of x.

For example, if $y = 3x + 1$, then for input $x = 2$, the output is $y = 7$.

Function notation allows you to keep track of both input and output. In function notation, the variable y is written as $f(x)$. This reminds you that the value of the function, $f(x)$, depends on the value of x.

Suppose $f(x) = 3x + 1$.

To find $f(2)$, substitute 2 for x in the equation.

$$f(2) = 3(2) + 1$$
$$f(2) = 7$$

Find $g(-3)$ if $g(x) = x^2 + 1$ Find $h(7)$ if $h(x) = (x - 5)^2$

$g(-3) = (-3)^2 + 1 = 9 + 1$ $h(7) = (7 - 5)^2 = (2)^2$

$g(-3) = 10$ $h(7) = 4$

Evaluate using function notation.

1. Let $f(x) = 3x + 1$. Find $f(0)$, $f(-1)$, and $f(1)$.

   $f(0) = 1, \ f(-1) = -2, \ f(1) = 4$

2. Let $g(x) = x^2 + 1$. Find $g(0)$, $g(1)$, and $g(5)$.

   $g(0) = 1, \ g(1) = 2, \ g(5) = 26$

3. Let $f(x) = (x - 5)^2$. Find $f(0)$, $f(1)$, and $f(-1)$.

   $f(0) = 25, \ f(1) = 16, \ f(-1) = 36$

4. Let $h(x) = 2x^2$. Find $h(0)$, $h(-1)$, and $h(2)$.

   $h(0) = 0, \ h(-1) = 2, \ h(2) = 8$

5. Let $f(x) = 8x - 11$. Find $f(0)$, $f(-3)$, and $f(5)$.

   $f(0) = -11, \ f(-3) = -35, \ f(5) = 29$

6. Let $g(x) = x^2 - 1$. Find $g(0)$, $g(1)$, and $g(-4)$.

   $g(0) = -1, \ g(1) = 0, \ g(-4) = 15$

7. Let $h(x) = (x + 1)^2$. Find $h(0)$, $h(1)$, and $h(-1)$.

   $h(0) = 1, \ h(1) = 4, \ h(-1) = 0$

31   **Holt Mathematics**

Problem Solving
Functions, Tables, and Graphs

Write the correct answer.

1. Film passes through a projector at the rate of 24 frames per second. The equation $y = 24x$ describes the number of frames, y, that have passed over any number of seconds, x. Complete the function table.

Input	Rule	Output
x	$24x$	y
2	$24 \cdot 2$	48
3	$24 \cdot 3$	72
4	$24 \cdot 4$	96
5	$24 \cdot 5$	120
6	$24 \cdot 6$	144

2. Anne pays \$40 a month for cable TV, plus \$4 for each movie she watches on pay-per-view channels. Complete the function table, where x is the number of pay-per-view movies she watches each month and y is her monthly cable bill.

Input	Rule	Output
x	$4x + 40$	y
3	$4(3) + 40$	52
5	$4(5) + 40$	60
7	$4(7) + 40$	68
9	$4(9) + 40$	76
11	$4(11) + 40$	84

Choose the letter of the best answer.

Madeline has a discount coupon for \$5 off her next purchase of tennis balls. Tennis balls are on sale for \$3 per can. The equation $y = 3x - 5$ gives her final cost, y, to purchase x cans of tennis balls.

3. If $x = 3$, what is the value of y?
   A 15   C 9
   B 14   (D) 4

4. If $x = 5$, what is the value of y?
   F 25   (H) 10
   G 15   J 5

5. If $x = 10$, what is the value of y?
   (A) 25   C 35
   B 30   D 50

6. If $x = 4$, what is the value of y?
   F 5   H 12
   (G) 7   J 17

7. If $x = 9$, what is the ordered pair (x, y)?
   A (5, 9)   (C) (9, 22)
   B (9, 5)   D (9, 10)

8. If $x = 6$, what is the ordered pair (x, y)?
   F (6, 18)   H (12, 31)
   G (6, 30)   (J) (6, 13)

32   **Holt Mathematics**

Reading Strategies
Focus On Vocabulary

A **function** is a special rule that pairs exactly one output value to each input value.

The value that replaces a variable is called an **input** value. The resulting value is called the **output**.

This equation shows the location of input and output in this function.

output   input
↓   ↓
$y = \quad 3x + 4$

This table lists input and output values for $y = 3x + 4$.

Input	Rule	Output
x	$3x + 4$	y
1	$3(1) + 4$	7
2	$3(2) + 4$	10
3	$3(3) + 4$	13
4	$3(4) + 4$	16

Answer the following questions.

1. Which variable stands for the input value?

   the variable x

2. Which variable stands for the output value?

   the variable y

3. What do you call the special relation or rule that exists between the input value and the output value?

   a function

4. An input value in a function can be paired with how many output values?

   one

5. In the table, which input value is paired with the output value 10?

   2

33   **Holt Mathematics**

Puzzles, Twisters & Teasers
Get to the Bottom of It!

Fill in the blanks to complete the function table.

Problem #	Input	Rule	Output	Ordered Pair
1	x	$2x$	y	(x,y)
2	-2	$2(-2)$	-4	$(-2,-4)$
3	-1	$2(-1)$	-2	$(-1,-2)$
4	0	$2(0)$	0	$(0,0)$
5	1	$2(1)$	2	$(1,2)$
6	2	$2(2)$	4	$(2,4)$
7	3	$2(3)$	6	$(3,6)$
8	4	$2(4)$	8	$(4,8)$
9	5	$2(5)$	10	$(5,10)$
10	6	$2(6)$	12	$(6,12)$

You've dropped your locker key in the pool and you must go through the maze to find it.

Step 1 – Go to start

Step 2 – Go right the number of the output in problem #9   **right 10**

Step 3 – Go down the number of the output in problem #7   **down 6**

Step 4 – Go left the absolute value of the output in problem #2   **left 4**

Step 5 – Go up the y value of the ordered pair in problem #6   **up 4**

Step 6 – Go right the x value of the ordered pair in problem #9   **right 5**

Sept 7 – Go down the y value in the ordered pair in problem #7   **down 6**

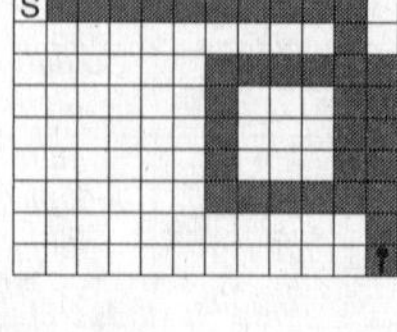

34   **Holt Mathematics**

          Holt Mathematics

Practice A
Find a Pattern in Sequences

Tell whether each sequence of y-values is arithmetic or geometric. Then find y when $n = 5$.

1.

n	1	2	3	4	5
y	3	6	12	24	■

geometric; 48

2.

n	1	2	3	4	5
y	7	12	17	22	■

arithmetic; 27

3.

n	1	2	3	4	5
y	12	30	48	66	■

arithmetic; 84

4.

n	1	2	3	4	5
y	4	12	36	108	■

geometric; 324

Choose the function from the box that describes each sequence.

$y = 4n$	$y = n - 4$	$y = n + 4$	$y = n - 1$	$y = n - 9$
$y = n + 6$	$y = n + 8$	$y = 8n$	$y = n - 0.1$	$y = 9n$

5. −3, −2, −1, 0, …

$y = n - 4$

6. 8, 16, 24, 32, …

$y = 8n$

7. 9, 10, 11, 12, …

$y = n + 8$

8. 0.9, 1.9, 2.9, 3.9, …

$y = n - 0.1$

9. 4, 8, 12, 16, …

$y = 4n$

10. 7, 8, 9, 10, …

$y = n + 6$

11. 0, 1, 2, 3, …

$y = n - 1$

12. 5, 6, 7, 8, …

$y = n + 4$

13. 9, 18, 27, 36, …

$y = 9n$

14. −8, −7, −6, −5, …

$y = n - 9$

Holt Mathematics

Practice B
Find a Pattern in Sequences

Tell whether each sequence of y-values is arithmetic or geometric. Then find y when $n = 5$.

1.

n	1	2	3	4	5
y	−5	10	25	40	■

arithmetic; 55

2.

n	1	2	3	4	5
y	14	28	56	112	■

geometric; 224

3.

n	1	2	3	4	5
y	4	12	36	108	■

geometric; 324

4.

n	1	2	3	4	5
y	28	42	56	70	■

arithmetic; 84

Write a function that describes each sequence.

5. 12, 24, 36, 48, …

$y = 12n$

6. 13, 14, 15, 16, …

$y = n + 12$

7. −8, −7, −6, −5, …

$y = n - 9$

8. 2.5, 3.5, 4.5, 5.5, …

$y = n + 1.5$

9. 9, 18, 27, 36, …

$y = 9n$

10. −12, −11, −10, −9, …

$y = n - 13$

11. $\frac{3}{4}$, $1\frac{3}{4}$, $2\frac{3}{4}$, $3\frac{3}{4}$, …

$y = n - \frac{1}{4}$

12. −3, −6, −9, −12, …

$y = -3n$

13. Mike ran 4 km on Sunday, 6 km on Monday, and 8 km on Tuesday. Write a function that describes the sequence. Then use the function to predict how many kilometers Mike will run on Friday.

$y = 2n + 2$; 14 km

Holt Mathematics

Practice C
Find a Pattern in Sequences

Tell whether each sequence of y-values is arithmetic or geometric. Then find y when $n = 5$.

1.

n	1	2	3	4	5
y	−3	9	−27	81	■

geometric; −243

2.

n	1	2	3	4	5
y	5	3.5	2	0.5	■

arithmetic; −1

Write a function that describes each sequence.

3. 7, 13, 19, 25, …

$y = 6n + 1$

4. 0.8, 1.2, 1.6, 2.0, …

$y = 0.4n + 0.4$

5. 1, 5, 9, 13, …

$y = 4n - 3$

6. −20, −15, −10, −5, …

$y = 5n - 25$

7. $\frac{4}{3}$, $\frac{5}{3}$, $\frac{6}{3}$, $\frac{7}{3}$, …

$y = \frac{1}{3}n + 1$

8. −7, −5, −3, −1, …

$y = 2n - 9$

Find a function that describes each sequence. Use the function to find the tenth term in the sequence.

9. −7, −3, 1, 5, …

$y = 4n - 11$; 29

10. 8, 17, 26, 35, …

$y = 9n - 1$; 89

11. 2.2, 3.2, 4.2, 5.2, …

$y = n + 1.2$; 11.2

12. −12, −6, 0, 6, …

$y = 6n - 18$; 42

13. Maria uses snap cubes to make a series of cubes. She makes the first cube by showing 1 snap cube. She makes the second cube by connecting 8 snap cubes, and the third cube by arranging 27 snap cubes. Write a function that describes this sequence. Then use the sequence to predict the number of snap cubes Maria will use to make the sixth cube.

$y = n^3$; 216 snap cubes

Holt Mathematics

Reteach
Find a Pattern in Sequences

Making a table is helpful for finding a function rule for a sequence of numbers.

Complete the tables for each sequence.

1. 4, 8, 12, 16, …

Term n	1	2	3	4	5	6	7
Value y	4	8	12	16	20	24	28

Rule: Add __4__ to each term to find the value of the next term.

Notice that the value of each term is 4 times the term number.

Term n	1	2	3	4	5	6	7
Rule	4 • 1	4 • 2	4 • 3	4 • 4	4 • 5	4 • 6	4 • 7
Value y	4	8	12	16	20	24	28

Write a function to describe this sequence.

Multiply the term number by __4__, or $y =$ __4__n.

2. 6, 7, 8, 9, …

Term n	1	2	3	4	5	6	7
Value y	6	7	8	9	10	11	12

Rule: Add __1__ to each term to find the value of the next term.

Notice that the value of each term is 5 more than the term number.

Term n	1	2	3	4	5	6	7
Rule	1 + 5	2 + 5	3 + 5	4 + 5	5 + 5	6 + 5	7 + 5
Value y	6	7	8	9	10	11	12

Write a function to describe this sequence.

Add __5__ to the term number, or $y = n +$ __5__

Write a function that describes each sequence.

3. 6, 12, 18, 24, …

$y = 6n$

4. 11, 12, 13, 14, …

$y = n + 10$

Holt Mathematics

Holt Mathematics

Challenge
LESSON 4-5 — Find the Term

You can use the pattern in a sequence to write an expression for the nth term of the sequence. Then you can use this expression to find any term in the sequence.

Find the fiftieth term of the following sequence: $\frac{1}{1}, \frac{1}{2}, \frac{1}{3}, \frac{1}{4}, \dots$

- The first term in this sequence is $\frac{1}{1}$. The numerator and the denominator of the fraction are both the same number as the location of the term in the sequence.

- The second term in this sequence is $\frac{1}{2}$. The numerator is 1, and the denominator is 2, the same number as the location of the term in the sequence.

- The third term in this sequence is $\frac{1}{3}$. The numerator is 1 and the denominator is 3, the same number as the location of the term in the sequence.

From the pattern, the nth term is $\frac{1}{n}$. So, the fiftieth term is $\frac{1}{50}$.

Solve.

1. Find the nth term and the twentieth term of the following sequence:
$-2, -1, 0, 1, \dots$

$n - 3, 17$

2. Find the nth term and the fortieth term of the following sequence:
$9, 10, 11, 12, \dots$

$n + 8, 48$

3. Find the nth term and the fifteenth term of the following sequence:
$5, 10, 15, 20, \dots$

$5n, 75$

4. Find the nth term and the twenty-fifth term of the following sequence:
$-2, -4, -6, -8, \dots$

$-2n, -50$

5. Find the nth term and the twentieth term of the following sequence: $\frac{1}{2}, \frac{2}{3}, \frac{3}{4}, \frac{4}{5}, \dots$

$\frac{n}{n+1}, \frac{20}{21}$

6. Find the nth term and the fiftieth term of the following sequence: $\frac{1}{2}, \frac{1}{4}, \frac{1}{6}, \frac{1}{8}, \dots$

$\frac{1}{2n}, \frac{1}{100}$

39

Problem Solving
LESSON 4-5 — Find a Pattern in Sequences

Write the correct answer.

1. Marina earns $15 for 1 hour of babysitting, $23 for 2 hours of babysitting, and $31 for 3 hours of babysitting. Write a function to describe the sequence.

$y = 8n + 7$

2. A website has 175 hits during its first hour of operation. It has 350 hits during its second hour and 525 hits during its third hour. Write a function to describe the sequence.

$y = 175n$

3. Rodney had 53 baseball cards. Then he started to buy cards each week. After 1 week, Rodney had 73 cards. After 2 weeks, he had 93 cards. After 3 weeks, he had 113 cards. Write a function to describe the sequence. Then use it to predict the number of cards Rodney will have after 5 weeks.

$y = 20n + 53; 153$ cards

4. Jen is reading a novel that is 463 pages long. After the first day, she had 423 pages left. After 2 days, she had 383 pages left. After 3 days, she had 343 pages left. Write a function to describe the sequence, and use it to predict the number of pages that Jen will have left after 8 days.

$y = 463 - 40n; 143$ pages

Choose the letter of the best answer.

Use the sequence 3, 6, 9, 12, … for the following exercises.

5. What is the rule for this sequence?
   A Add 3 to n.
   B Subtract 3 from n.
   C Multiply n by 3.
   D Divide n by 3.

6. What is the function that describes this sequence?
   F $y = 3n$
   G $y = \frac{n}{3}$
   H $y = n + 3$
   J $y = n - 3$

7. What are the next 3 terms in the sequence?
   A 9, 6, 3
   B 15, 18, 21
   C 15, 19, 24
   D 36, 432, 5,184

8. What is the ninth term in the sequence?
   F 24
   G 27
   H 30
   J 33

40

Reading Strategies
LESSON 4-5 — Analyze Information

A **sequence** is a list of values arranged in a special order or pattern. Numbering the terms in a sequence helps you analyze the pattern.

Term	1st	2nd	3rd	4th
Value	3	6	9	12

1. What do you call a list of numbers that create a pattern?

a sequence

2. What is the value of the first term in the sequence above?

3

3. The number 9 is the value for which term in the sequence?

the third term

4. What pattern can you see for the values in the table?

The values increase by 3.

A function table can help you see the relation between the number of the term and value of the term.

Number of term: n	Rule	Value of term: y
1	$1 \cdot 3$	3
2	$2 \cdot 3$	6
3	$3 \cdot 3$	9
4	$4 \cdot 3$	12

Study the function table to complete each question.

5. When the value of n is 2, what is the value of y?

6

6. When the value of n is 4, what is the value of y?

12

7. What pattern do you see for the value of y?

The number of the term is multiplied by 3 to get y.

8. What would be the values of the fifth term and the sixth term? ___ 15 and 18

41

Puzzles, Twisters, & Teasers
LESSON 4-5 — Weigh the Options!

Decide whether each sequence of y-values is arithmetic or geometric. Circle the letter above your answer.

1.

n	1	2	3	4
y	10	20	30	40

T — arithmetic   A — geometric

2.

n	1	2	3	4
y	7	21	63	189

F — arithmetic   H — geometric

3.

n	1	2	3	4
y	2	8	32	128

U — arithmetic   I — geometric

4.

n	1	2	3	4
y	36	30	24	18

R — arithmetic   S — geometric

Next, choose the function that describes each sequence. Circle the letter above your answer.

5. $-5, -4, -3, -2, \dots$
   L — $y = n - 5$ W — $y = n - 6$

6. $3.5, 4.5, 5.5, 6.5, \dots$
   D — $y = 3.5n$ N — $y = n + 2.5$

7. $30, 60, 90, 120, \dots$
   C — $y = 30n$ H — $y = n + 30$

8. $\frac{1}{4}, 1\frac{1}{4}, 2\frac{1}{4}, \dots$
   A — $y = n - \frac{3}{4}$ O — $y = n + \frac{1}{4}$

9. $-9, -8, -7, -6, \dots$
   T — $y = -9n$ E — $y = n - 10$

10. $25, 50, 75, 100, \dots$
    S — $y = 25n$ D — $y = n + 25$

Start with problem number 1. Write the circled letters above the problem numbers to solve the riddle.

Why are fish easy to weigh?

They have

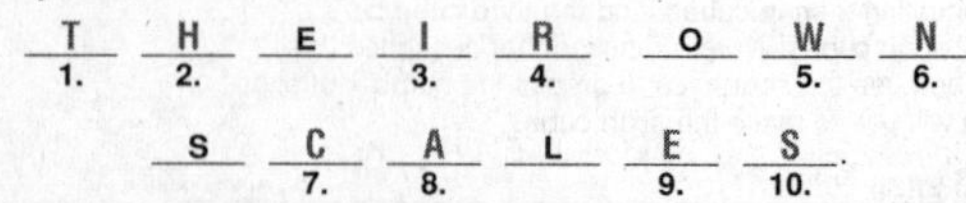

$\underset{\text{1.}}{T}\ \underset{\text{2.}}{H}\ \underset{\text{3.}}{E}\ \underset{\text{?}}{I}\ \underset{\text{4.}}{R}\ \underset{}{O}\ \underset{\text{5.}}{W}\ \underset{\text{6.}}{N}$

$\underset{\text{7.}}{S}\ \underset{\text{8.}}{C}\ \underset{}{A}\ \underset{}{L}\ \underset{\text{9.}}{E}\ \underset{\text{10.}}{S}$

42

60

Practice A
Graphing Linear Functions

Complete the function tables. Then match the letter of each graph with the function table for its linear function.

1. $y = x - 2$ Graph: __C__

Input	Linear Equation	Output	Ordered Pair
x	$y = x - 2$	y	(x, y)
0	$y = 0 - 2$	-2	$(0, -2)$
1	$y = 1 - 2$	-1	$(1, -1)$
2	$y = 2 - 2$	0	$(2, 0)$

A
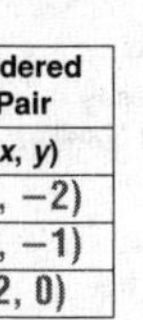

2. $y = \frac{x}{2}$ Graph: __A__

Input	Linear Equation	Output	Ordered Pair
x	$y = \frac{x}{2}$	y	(x, y)
0	$y = \frac{0}{2}$	0	$(0, 0)$
2	$y = \frac{2}{2}$	1	$(2, 1)$
4	$y = \frac{4}{2}$	2	$(4, 2)$

B
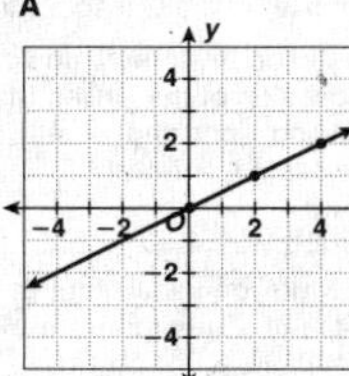

3. $y = -2$ Graph: __B__

Input	Linear Equation	Output	Ordered Pair
x	$y = -2$	y	(x, y)
0	$y = -2$	-2	$(0, -2)$
1	$y = -2$	-2	$(1, -2)$
2	$y = -2$	-2	$(2, -2)$

C
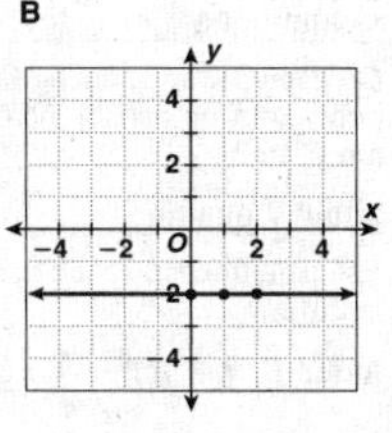

Holt Mathematics

Practice B
Graphing Linear Functions

Graph each linear function.

1. $y = -x - 5$

Input	Linear Equation	Output	Ordered Pair
x	$y = -x - 5$	y	(x, y)
-4	$y = -(-4) - 5$	-1	$(-4, -1)$
-2	$y = -(-2) - 5$	-3	$(-2, -3)$
0	$y = -0 - 5$	-5	$(0, -5)$

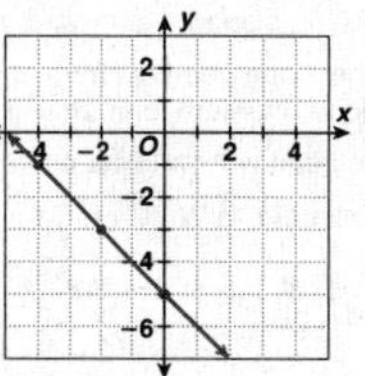

2. $y = 2x - 1$

Input	Linear Equation	Output	Ordered Pair
x	$y = 2x - 1$	y	(x, y)
-2	$y = 2(-2) - 1$	-5	$(-2, -5)$
0	$y = 2(0) - 1$	-1	$(0, -1)$
1	$y = 2(1) - 1$	1	$(1, 1)$

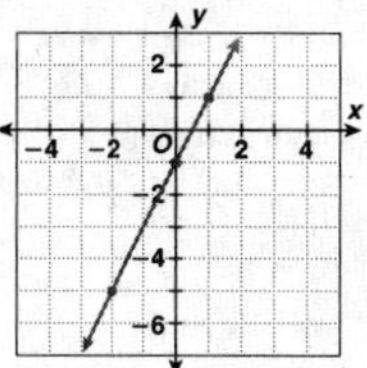

3. The temperature of a swimming pool is 75°F. When the pool heater is turned on, the temperature rises 2°F every hour. What will the temperature be after 3 hours? Make a function table to answer the question.

__81°F__

Input	Equation	Output
x	$y = 2x + 75$	y
1	$y = 2(1) + 75$	77
2	$y = 2(2) + 75$	79
3	$y = 2(3) + 75$	81

4. Mel's Pizza Place charges $15.00 for a large cheese pizza plus $1.25 for each additional topping. What will be the cost of a large pizza with 3 additional toppings? Make a function table to answer the question.

__$18.75__

Input	Equation	Output
x	$y = 1.25x + 15$	y
1	$y = 1.25(1) + 15$	16.25
2	$y = 1.25(2) + 15$	17.50
3	$y = 1.25(3) + 15$	18.75

Holt Mathematics

Practice C
Graphing Linear Functions

Graph each linear function.

1. $y = x - 4$

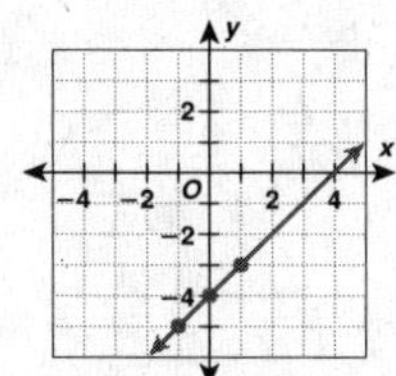

2. $y = \frac{x}{3} - 3$

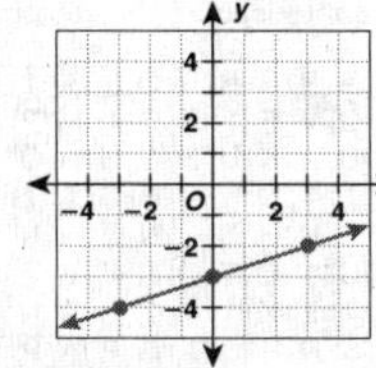

3. $y = x + 3$

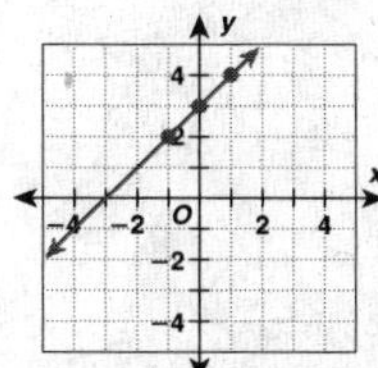

4. $y = 3x + 4$

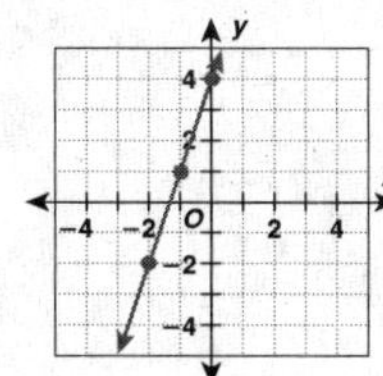

5. Jackie is saving her money to buy a coat that costs $121. If she already has $61 and saves $15 each week, in how many weeks can she buy the coat? Make a function table to answer the question.

__4 weeks__

Input	Equation	Output
x	$y = 15x + 61$	y
1	$y = 15(1) + 61$	76
2	$y = 15(2) + 61$	91
3	$y = 15(3) + 61$	106
4	$y = 15(4) + 61$	121

Holt Mathematics

Reteach
Graphing Linear Functions

The graph of a linear equation is a straight line. A **linear function** is a function whose graph is a straight line that is not vertical. To graph a linear equation, first make a function table.

Complete the function table for $y = 2x - 3$.

	Input	Linear Equation	Output	Ordered Pair
	x	$y = 2x - 3$	y	(x, y)
1.	0	$y = 2(0) - 3 = 0 - 3$	-3	$(0, -3)$
2.	1	$y = 2(1) - 3 = 2 - 3$	-1	$(1, -1)$
3.	2	$y = 2(2) - 3 = 4 - 3$	1	$(2, 1)$

To graph the equation, graph ordered pairs. Then draw a line through the points.

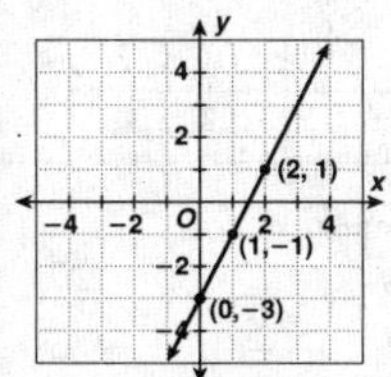

Complete the function table and graph the equation.

4. $y = -x + 1$

Input	Linear Equation	Output	Ordered Pair
x	$y = -x + 1$	y	(x, y)
0	$y = -0 + 1$	1	$(0, 1)$
1	$y = -1 + 1$	0	$(1, 0)$
2	$y = -2 + 1$	-1	$(2, -1)$

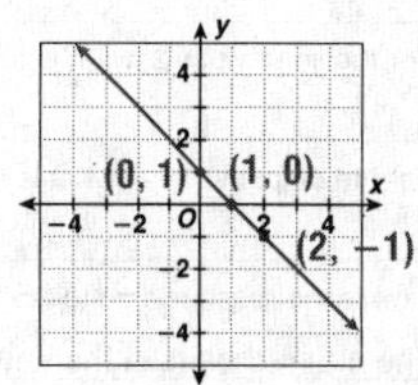

Holt Mathematics

Holt Mathematics

Challenge
Linear Functions

A linear equation is referred to as a "first-degree polynomial equation." The *standard form* of a first-degree equation is $ax + by + c = 0$. In standard form, a, b, and c are integers.

The *slope-intercept form* of a linear equation is $y = mx + b$. In slope-intercept form, m and b are both real numbers.

You can write a linear equation in either of these forms.

Example 1: Write $y = \frac{2}{3}x - 2$ in standard form.

$$y = \frac{2}{3}x - 2$$
$$y + 2 = \frac{2}{3}x \quad \longleftarrow \text{Add 2 to both sides.}$$
$$3(y + 2) = 3\left(\frac{2}{3}x\right) \quad \longleftarrow \text{Multiply both sides by 3.}$$
$$3y + 6 = 2x \quad \longleftarrow \text{Simplify.}$$
$$-2x + 3y + 6 = 0 \quad \longleftarrow \text{Subtract } 2x \text{ from both sides.}$$

Example 2: Write $3x + 5y + 1 = 0$ in slope-intercept form.
$$3x + 5y + 1 = 0$$
$$5y = -3x - 1 \quad \longleftarrow \text{Subtract } 3x + 1 \text{ from both sides.}$$
$$\frac{5y}{5} = \frac{-3x - 1}{5} \quad \longleftarrow \text{Divide both sides by 5.}$$
$$y = -\frac{3}{5}x - \frac{1}{5} \quad \longleftarrow \text{Simplify.}$$

Write each equation in standard form.

1. $y = \frac{5}{6}x - 4$

2. $y = 2x - \frac{1}{2}$

3. $y = -\frac{1}{3}x + 8$

$-5x + 6y + 24 = 0$ $-4x + 2y + 1 = 0$ $x + 3y - 24 = 0$

or $5x - 6y - 24 = 0$ or $4x - 2y - 1 = 0$ or $-x - 3y + 24 = 0$

Write each equation in slope-intercept form.

4. $x - 4y + 1 = 0$

5. $2x - 4y + 8 = 0$

6. $6x - 3y - 2 = 0$

$y = \frac{1}{4}x + \frac{1}{4}$ $y = \frac{1}{2}x + 2$ $y = 2x - \frac{2}{3}$

47   **Holt Mathematics**

Problem Solving
Graphing Linear Functions

Write the correct answer.

This graph shows the approximate population density (people per square mile) in the United States from 1950 to 2000.

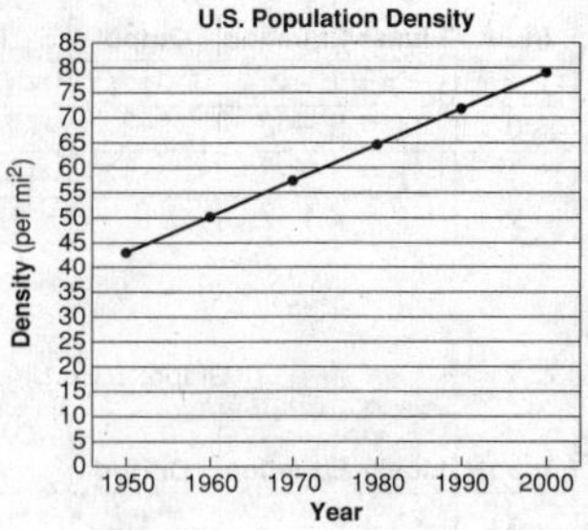

1. Based on this graph, do you think the population of the United States is growing, shrinking, or staying about the same?

   growing

2. How do you know that this is the graph of a linear function?

   because the graph is a

   straight line

3. On average, by about how many people did the density increase every ten years?

   about 7 people

U.S. Population Density

(Graph: Density (per mi²) on y-axis from 0 to 85, Year on x-axis from 1950 to 2000, showing a rising straight line.)

4. Estimate the population density in 2010.

   about 87 per mi²

5. Estimate the population density in 2050.

   about 115 per mi²

Choose the letter of the best answer.

6. Given the linear equation $y = 4x + 1$ and the input $x = 3$, what is the resulting ordered pair?

   A (15, 3)   C (12, 3)
   B (3, 13)   D (12, 11)

7. Given the linear equation $y = \frac{1}{3}x - 4$ and the input $x = 6$, what is the resulting ordered pair?

   F (6, −2)   H (−3, 6)
   G (6, 2)   J $\left(\frac{1}{3}, 6\right)$

8. A technician is adding chemicals to a 210-liter tank. She is adding the liquid at a rate of 2.5 liters per minute. How long will it take to fill the tank half full?

   A 42 min   C 168 min
   B 84 min   D 525 min

9. A basic set of 4 golf lessons costs $95.00. Additional lessons can be purchased at the discounted rate of $15.00 each. What will Veronica pay for a series of 10 lessons?

   F $150   **H $185**
   G $155   J $245

48   **Holt Mathematics**

Reading Strategies
Use a Graphic Organizer

This chart can help you understand the different ways to represent a linear function.

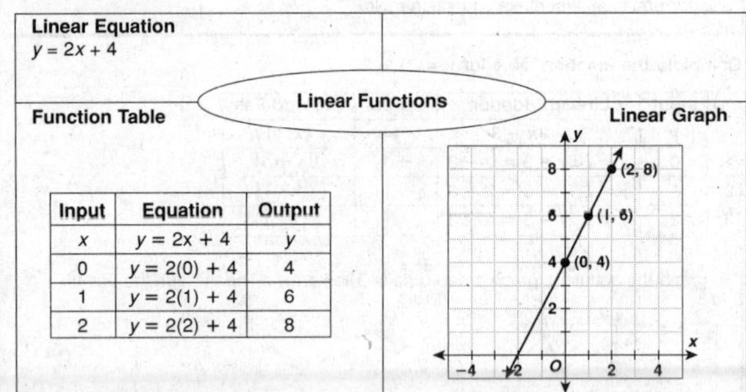

Input	Equation	Output
x	$y = 2x + 4$	y
0	$y = 2(0) + 4$	4
1	$y = 2(1) + 4$	6
2	$y = 2(2) + 4$	8

(Graph showing points (0, 4), (1, 6), (2, 8))

Use the graphic organizer to answer the following questions.

1. In addition to a linear equation, what are two other ways to represent a linear function?

   a graph and a function table

2. Write the linear equation for the graph above.

   $y = 2x + 4$

3. For the input value 2, what is the output value?

   8

4. For the output value 4, what is the input value?

   0

5. How is the graph related to the function table?

   The ordered pairs of the points in the graph are the pairs of input and output values in the table.

49   **Holt Mathematics**

Puzzles, Twisters & Teasers
Graduation Day!

Find and circle words from the list in the word search. Find a word that solves the riddle. Circle it and write it on the line.

linear   equation   function   graph   line
representation   input   output   point   domain

```
L I N E A R G B D O M A I N
I B N M O V F F E P L M P K
N S C D U I K U E Y H N O I
E Q U A T I O N P B G T I F
I C F V P N H C L N J I N B
N N J O U I L T O R F V T D
P M K P T Z X I M T G B N J
U B H I K N K O A G R A P H
T M J T U H B N W D F R B H
R E P R E S E N T A T I O N
Q W E R T Y U I O P L K J H
```

What does a fish get when it graduates from school?

A D E E P L O M A

50   **Holt Mathematics**

62

Holt Mathematics